Applied Condition Monitoring

Volume 12

The book series Applied Condition Monitoring publishes the latest research and developments in the field of condition monitoring, with a special focus on industrial applications. It covers both theoretical and experimental approaches, as well as a range of monitoring conditioning techniques and new trends and challenges in the field. Topics of interest include, but are not limited to: vibration measurement and analysis; infrared thermography; oil analysis and tribology; acoustic emissions and ultrasonics; and motor current analysis. Books published in the series deal with root cause analysis, failure and degradation scenarios, proactive and predictive techniques, and many other aspects related to condition monitoring. Applications concern different industrial sectors: automotive engineering, power engineering, civil engineering, geoengineering, bioengineering, etc. The series publishes monographs, edited books, and selected conference proceedings, as well as textbooks for advanced students.

More information about this series at http://www.springer.com/series/13418

Ahmed Felkaoui · Fakher Chaari
Mohamed Haddar
Editors

Rotating Machinery and Signal Processing

Proceedings of the First Workshop on Signal Processing Applied to Rotating Machinery Diagnostics, SIGPROMD'2017, April 09–11, 2017, Setif, Algeria

Springer

Editors
Ahmed Felkaoui
Institute of Optics and Precision Mechanics
University Ferhat Abbas
Sétif, Algeria

Mohamed Haddar
National School of Engineers of Sfax
Sfax, Tunisia

Fakher Chaari
National School of Engineers of Sfax
Sfax, Tunisia

ISSN 2363-698X ISSN 2363-6998 (electronic)
Applied Condition Monitoring
ISBN 978-3-030-07149-3 ISBN 978-3-319-96181-1 (eBook)
https://doi.org/10.1007/978-3-319-96181-1

This Springer imprint is published by the registered company Springer Nature Switzerland AG
The registered company address is: Gewerbestrasse 11, 6330 Cham, Switzerland

Preface

The first workshop on Signal Processing Applied to Rotating Machinery Diagnostics (SIGPROMD'2017) was held in Setif, Algeria, in April 2017. This event was organized jointly by the Applied Precision Mechanics Laboratory (LMPA) of the Institute of Precision Mechanics, University of Setif, Algeria; and the Mechanics, Modeling and Manufacturing Laboratory (LA2MP) of the National School of Engineers of Sfax, Tunisia.

All the chapters included in this book were rigorously reviewed by two referees. Our thanks go to all reviewers of the 12 papers composing this proceeding published under Applied Condition Monitoring book series.

It is well known that rotating machinery gives rise to vibrations and consequently noise. Vibration signature depends on the setting up and the health status of each machine. A change in the vibration signature induced by a change in the machine state is a powerful mean to detect incipient defects before they evolve and become critical. Vibration signals collected from machines should be processed in order to extract state features which are compared to reference values. The objective of the workshop was to gather researchers from both laboratories to discuss latest advances in signal processing dedicated to rotating machinery. It was a forum to exchange ideas and developments in this field. The main topics that were discussed during the workshop through the presented chapters are:

- Noise and vibration of machines
- Condition monitoring in non-stationary operations
- Vibro-acoustic diagnosis of machinery
- Signal processing
- Pattern recognition
- Monitoring and diagnostic systems
- Modeling of dynamics and faults in machinery

The editors would like to thank all participants in SIGPROMD'2017 for their valuable contribution to this book. They hope that the readers can find what they expect in the field of signal processing dedicated to machinery diagnostics. Finally, many thanks go to Springer for offering this opportunity to publish the proceedings of the workshop.

Setif/Sfax
2018

Ahmed Felkaoui
Fakher Chaari
Mohamed Haddar

Contents

Feature Selection Scheme Based on Pareto Method for Gearbox Fault Diagnosis 1
Ridha Ziani, Hafida Mahgoun, Semcheddine Fedala, and Ahmed Felkaoui

Intelligent Gear Fault Diagnosis in Normal and Non-stationary Conditions Based on Instantaneous Angular Speed, Differential Evolution and Multi-class Support Vector Machine 16
Semchedine Fedala, Didier Rémond, Ahmed Felkaoui, and Houssem Selmani

Effect of Input Data on the Neural Networks Performance Applied in Bearing Fault Diagnosis 34
Hocine Fenineche, Ahmed Felkaoui, and Ali Rezig

Bearing Diagnostics Using Time-Frequency Filtering and EEMD 44
Hafida Mahgoun and Ridha Ziani

The Time-Frequency Filtering (TFF) Method Used in Early Detection of Gear Faults in Variable Load and Dimensions Defect 56
Hafida Mahgoun, Fakher Chaari, Ahmed Felkaoui, and Mohamed Haddar

Comparison Between Hidden Markov Models and Artificial Neural Networks in the Classification of Bearing Defects 68
Miloud Sedira, Ridha Ziani, and Ahmed Felkaoui

On-line Adaptive Scaling Parameter in Active Disturbance Rejection Controller 79
Maroua Haddar, S. Caglar Baslamisli, Fakher Chaari, and Mohamed Haddar

Modal Analysis of the Clutch Single Spur Gear Stage System with Eccentricity Defect 87
Ahmed Ghorbel, Moez Abdennadher, Lassâad Walha, Becem Zghal, and Mohamed Haddar

Estimation of Road Disturbance for a Non Linear Half Car Model Using the Independent Component Analysis 96
Dorra Ben Hassen, Mariem Miladi, Mohamed Slim Abbes, S. Caglar Baslamisli, Fakher Chaari, and Mohamed Haddar

Transfer Path Analysis of Planetary Gear with Mechanical Power Recirculation 104
Ahmed Hammami, Alfonso Fernandez del Rincon, Fakher Chaari, Fernando Viadero Rueda, and Mohamed Haddar

Modeling the Transmission Path Effect in a Planetary Gearbox 116
Oussama Graja, Bacem Zghal, Kajetan Dziedziech, Fakher Chaari, Adam Jablonski, Tomasz Barszcz, and Mohamed Haddar

Dynamic Behavior of Spur Gearbox with Elastic Coupling in the Presence of Eccentricity Defect Under Acyclism Regime 123
Atef Hmida, Ahmed Hammami, Fakher Chaari, Mohamed Taoufik Khabou, and Mohamed Haddar

Author Index 133

Feature Selection Scheme Based on Pareto Method for Gearbox Fault Diagnosis

Ridha Ziani(✉), Hafida Mahgoun, Semcheddine Fedala, and Ahmed Felkaoui

Laboratory of Applied Precision Mechanics, Institute of Optics and Precision Mechanics, Ferhat Abbes University, Setif 1, 19000 Setif, Algeria
ziani_lmpa@yahoo.fr

Abstract. Fault diagnosis based on pattern recognition approach has three main steps viz. feature extraction, sensitive features selection, and classification. The vibration signals acquired from the system under study are processed for feature extraction using different signal processing methods. Followed by feature selection process, classification is performed. The challenge is to find good features that discriminate the different fault conditions of the system, and increase the classification accuracy. This paper proposes the use of Pareto method for optimal feature subset selection from the pool of features. To demonstrate the efficiency and effectiveness of the proposed fault diagnosis scheme, numerical analyses have been performed using the Westland data set. The Westland data set consists of vibration data collected from a US Navy CH-46E helicopter gearbox in healthy and faulty conditions. First, features are extracted from vibration signals in time, spectral, and time-scale domain, then ranked according to three different criterions namely: Fisher score, correlation, and Signal to Noise Ratio (SNR). Afterword, data formed by only the selected features is used as input for the classification problem. The classification task is achieved using Support Vector Machines (SVM) method. The proposed fault diagnosis scheme has shown promising results. Using only the feature subset selected by Pareto method with Fisher criterion, SVMs achieved 100% correct classification.

Keywords: Signal processing · Support Vector Machine · Feature selection Vibration · Fault diagnosis

1 Introduction

The gears are one of the major components of rotating machines, and proper maintenance of gear system is very essential to ensure reliability, safety, and performance of machines. The most of the developed methods for fault diagnosis of these systems are based on pattern recognition approach (Rafiee et al. 2007, 2010; Gryllias and Antoniadis 2012; Zhang et al. 2013; Ziani et al. 2017). The advantage of this approach is that it doesn't require large priori knowledge of the process under study. In this case, the diagnosis is assimilated to a classification problem (healthy or faulty condition).

A. Felkaoui et al. (Eds.): SIGPROMD'2017, ACM 12, pp. 1–15, 2019.
https://doi.org/10.1007/978-3-319-96181-1_1

The specialty of condition features is to provide accurate information regarding the condition of various components at different levels of damage (initial, heavy, or growing).

Vibration analysis is considered as a the most suitable tool for rotating machines faults diagnosis, thus it has attracted greater attention towards the researchers to acquire, analyze and quantify this parameter for improving the diagnosis precision. A multitude of methods have been developed. The yield of these techniques is, to distinguish changes in the signal brought on because of damaged or faulty components. These techniques are generally based on signal processing in different domains: time, spectral, time-frequency, and time-scale.

In time domain, the analysis is generally based on statistical features which provide an overall picture of some aspect of the time-series under investigation. Examples of these features include arithmetic mean, root mean square (RMS), variance (or standard deviation), skewness, kurtosis, peak-to-peak, crest factor (Ziani et al. 2017). Time Synchronous Averaging (TSA) (Abdul Rahman et al. 2011) is a pre-processing technique which was widely used for signal denoising before performing the feature extraction procedure.

In frequency domain, the most popular technique is Fast Fourier Transform (FFT) which provides a representation of the frequency content of a given signal. Various techniques resulted from FFT such as Power Spectral Density (PSD), cestrum analysis, and envelope analysis. Many authors used amplitudes, entropy, and significant energy, calculated around fault characteristic frequency, to form the feature vector.

Time frequency distributions represent a good way to analyze the non stationary mechanical signals in which the spectral content changes with time. Short Time Fourier Transform (STFT), and Wigner–Ville distribution (Baydar and Ball 2001) are the well known time frequency distributions employed to overcome this problem, and widely used to processing the vibration signals of systems operating in non stationary modes.

The non-stationary signals can be considered as a superposition of components with respect to a set of basis functions which are each more or less localized in time. These basis functions can then be used to represent different frequency content simply by scaling them with respect to time. Signal decomposition using such called functions results in the so-called time-scale representations—and this leads directly to the wavelet transform (Worden et al. 2011).

Another group of features which have grown in popularity in recent years are those based on the Empirical Mode Decomposition (EMD) and Hilbert–Huang transform (HHT). These nonlinear analysis methods were employed to deal with the non-stationary vibrations to extract the original fault feature vector (Mahgoun et al. 2016).

A review on the application of the above signal processing methods and others for gear fault diagnosis can be found in Goyal et al. (2016).

In fault diagnosis methods based on pattern recognition approach, irrelevant features spoil the performance of the classifier and reduce the recognition accuracy (Kudo and Sklansky 2000). Hence it is necessary to reduce the dimension of the data by finding a small set of important features which can give good classification performance.

Dimensionality reduction is one of the most popular techniques to remove irrelevant and redundant features. Dimensionality reduction techniques may be divided in two main categories, called feature extraction (FE) and feature selection (FS) (Kotsiantis 2011). Feature extraction approaches map the original feature space to a new feature space with lower dimensions by combining the original feature space. This transformation may be a linear or nonlinear combination of the original features. These methods include Principle Component Analysis (PCA), Linear Discriminant Analysis (LDA) and Canonical Correlation Analysis (CCA). Bartkowiak and Zimroz (2014) cited other transformation methods used for reducing the dimensionality of the data, such as: Independent Component Analysis (ICA), Isomap, local linear embedding, kernel PCA, and curvilinear component analysis. On the other hand the term feature selection refers to algorithms that output a subset of the input feature set.

Both Feature extraction and feature selection are capable of improving learning performance, lowering computational complexity, building better generalizable models, and decreasing required storage (Tang et al. 2014). While feature selection selects a subset of features from the original feature set without any transformation, and maintains the physical meanings of the original features, it is better to select and process original data than create new features because by projections the physical meaning of the original variables may be lost (Bartkowiak and Zimroz 2014).

For the classification problem, algorithms used to select features are divided into three categories: filter, wrapper, and embedded methods (Tang et al. 2014). Filter methods rank features or feature subsets independently of the classifier, while wrapper methods use the predictive accuracy of a classifier to assess feature subsets, thus, these methods are usually computationally heavy and they are conditioned to the type of classifier used. Another type of feature subset selection is identified as embedded methods. In this case, the feature selection process is done inside the induction algorithm itself, i.e. attempting to jointly or simultaneously train both a classifier and a feature subset. They often optimize an objective function that jointly rewards the accuracy of classification and penalizes the use of more features (Kotsiantis 2011).

The goal of this study is to present a feature selection scheme based on Pareto method combined with three different criterions namely: Fisher score, Correlation criterion, and Signal to Noise Ratio (SNR). This approach was tested using vibration data acquired from a helicopter gearbox. In this study, Support Vector Machines (SVM) was used to achieve the classification task. This method has a good generalization capability even in the small-sample cases of classification and has been successfully applied in fault detection and diagnosis in Gryllias and Antoniadis (2012), Ziani et al. (2017), Konar and Chattopadhyay (2011).

The rest of this paper is organized as follow: In the second section we present the basic principle of SVM. Vibration data and feature extraction procedure are given the third section. In the fourth section we present the proposed feature selection method. Results are presented and discussed in the fifth section. Finally, the sixth section is dedicated to the conclusion.

2 Support Vector Machines (SVMs)

SVMs is a relatively a new computational learning method proposed by Vapnik (1998). The essential idea of SVMs is to place a linear boundary between two classes of data, and adjust it in such a way that the margin is maximized, namely, the distance between the boundary and the nearest data point in each class is maximal. The nearest data points are known as Support Vectors (SVs) (Konar and Chattopadhyay 2011). Once the support vectors are selected, all the necessary information to define the classifier is provided.

If the training data are not linearly separable in the input space, it is possible to create a hyper plane that allows linear separation in the higher dimension. This is achieved through the use of a transformation that converts the data from an N-dimensional input space to Q-dimensional feature space. A kernel can be used to perform this transformation. Among the kernel functions in common use are linear functions, polynomials functions, Radial Basis Functions (RBF), and sigmoid functions. A deeper mathematical treatise of SVMs can be found in the book of Vapnik (1998) and the tutorials on SVMs (Burges 1998; Scholkopf 1998).

SVMs is essentially a two-class classification technique, which has to be modified to handle the multiclass tasks in real applications e.g. rotating machinery which usually suffer from more than two faults. Two of the common methods to enable this adaptation include the One-against-all (OAA) and One-against-one (OAO) strategies (Yang et al. 2005).

In the One-against-all strategy, each class is trained against the remaining N − 1 classes that have been collected together. The "winner-takes-all" rule is used for the final decision, where the winning class is the one corresponding to the SVM with the highest output (discriminant function value). For one classification, N two-class SVMs are needed.

The One-against-one strategy needs to train N (N − 1)/2 two-class SVMs, where each one is trained using the data collected from two classes. When testing, for each class, score will be computed by a score function. Then, the unlabeled sample x will be associated with the class with the largest score.

3 Vibration Data and Feature Extraction

3.1 The CH46 Gearbox

Vibration data used in this paper is acquired from the Westland CH46 Helicopter gearbox (Cameron 1993). The gearbox is relatively complex, driving both the main shaft and many auxiliary devices. This vibration data have been widely used to validate the effectiveness of new algorithms for gear fault diagnosis (Williams and Zalubas 2000; Loughlin and Cakrak 2000; Chang et al. 2009; Nandi et al. 2013).

Figure 1 shows the simplified main section of the CH46 helicopter gearbox including the input, quill and output shafts, the spur pinion/collector gear pair and the spiral bevel pinion/gear pair. In this study we interest only to fault of gear 5, (spiral bevel pinion tooth spalling). For this element, the vibration data is composed of twenty four (24) signals: nine (9) signals acquired in normal condition (Fig. 2a), six (6) with defect Level 1 condition (Fig. 2b), and nine (9) with defect level 2 (Fig. 2c).

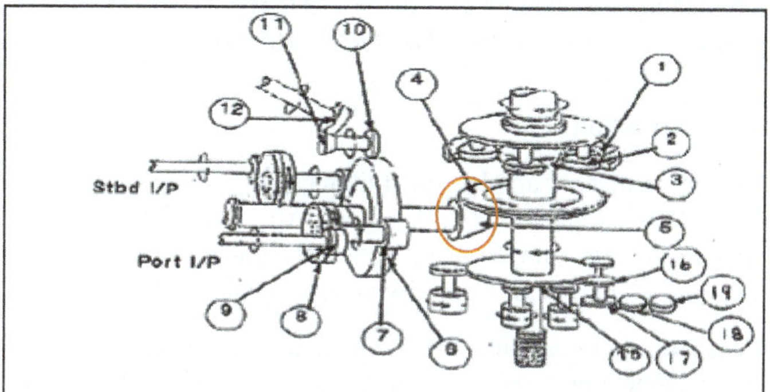

GEAR N°	DESCRIPTION	N° OF TEETH	MESH FREQUENCY (HZ)
1	RING GEAR	117	514.8
2	PLANET GEAR	39	514.8
3	SUN GEAR	39	514.8
4	SPIRAL BEVEL GEAR	63	1108.9
5	SPIRAL BEVEL PIGNON	26	1108.9
6	COLLECTOR GEAR	74	3155.8
7	SPUR PIGNON	25	3155.8
8	MIX BOX HELICAL IDLER GEAR	72	9088.8
9	MIX BOX INPUT PIGNON	28	9088.8
10	BLOWER SPUR GEAR	25	3155.8
11	BLOWER BEVEL GEAR	25	3155.8
12	BLOWER PIGNON	31	3155.8
15	AUX DRIVE SPUR GEAR	130	2288
16	ACCESSORY DRIVE AFT	20	2288
17	ROTOR POSITION DRIVE	42	4804.8
18	SCAVANGE PUMP DRIVE	62	4804.8
19	OPTICAL TACH DRIVE	42	4804.8

Fig. 1. Simplified main section of the CH46 helicopter gearbox

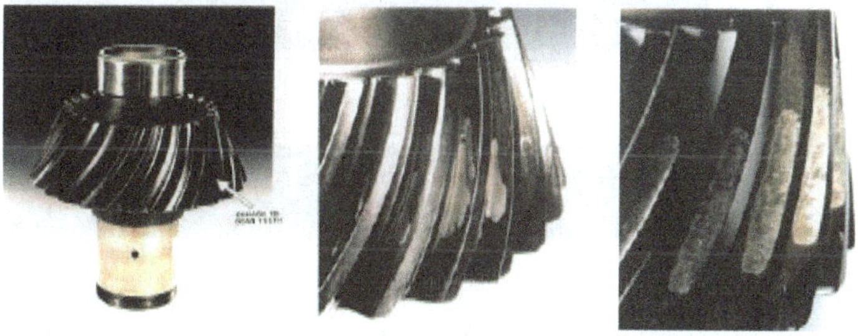

a) General view of spiral bevel input pinion (Normal)

b) Spiral bevel input pinion tooth spalling (Defect level 1)

c) Spiral bevel input pinion tooth spalling (Defect level 2)

Fig. 2. The Bevel input pinion used in the test

Signals are composed of 412464 samples acquired with a sampling frequency of 103116 Hz. The following parameters are given:

- Number of teeth of spiral bevel pinion/gear: n1 = 26; n2 = 63;
- Rotating frequency $fr1$ = 42.65 Hz; $fr2$ = 17.60 Hz;
- Meshing frequency: $fm1$ = 1108.9 Hz, $fm2$ = 3155 Hz.

3.2 Features Extraction

In order to obtain sufficient samples for training and testing SVMs, each signal was divided into ten (10) samples of 41246 points. Afterwards, different feature subsets were extracted from each sample using different signal processing methods. These features were extracted in time domain, frequency domain, and time frequency domain.

Statistical Features. In time domain (Fig. 3), signals are processed to extract the nine following statistical features: mean, Root Mean Square (RMS), skewness, kurtosis, Peak factor, Peak to Peak value, Clearance factor, Shape factor, and Impulse factor. The mathematical formula of these features can be found in Goyal et al. (2016).

Spectral Features. In spectral domain, another feature subset is formed by calculating the Power Spectral Density (PSD) in different bands around the meshing frequency and its four harmonics. The width of each band is chosen equal to ten rotating frequency (426 Hz). Consequently the frequency bands are: [895–1321 Hz], [2004–2430 Hz], [3113–3539 Hz] [4222–4648 Hz], and [5331–5757 Hz].

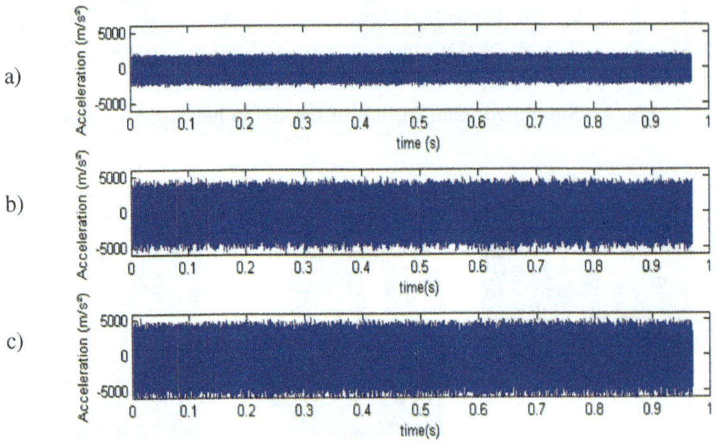

Fig. 3. Times domain signals acquired under torque of 45%. (a) Normal condition (b) defect level 1, (c) defect level 2

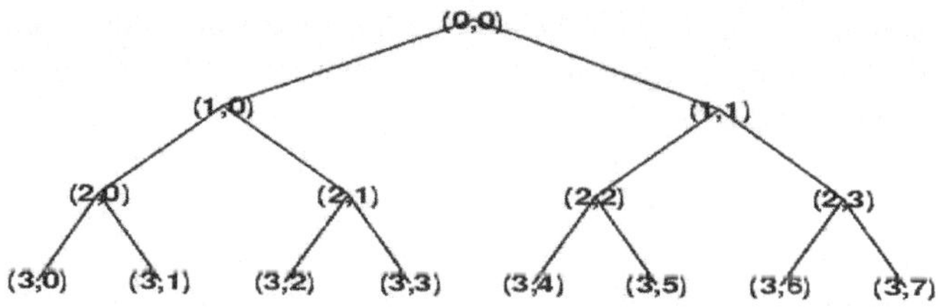

Fig. 4. Wavelet packet decomposition tree

Wavelet Packet Decomposition. In the last decade, Wavelet Packet Decomposition (WPD) has been proved to be a suitable tool for gear fault diagnosis, especially on vibration signal features extraction (Zhang et al. 2013). WPD shows good performance on both high and low frequency analysis. The selection of the mother wavelet is a crucial step in wavelet analysis. In (Rafiee et al. 2010) it has been shown that the Daubechies 44 wavelet is the most effective for both faulty gears and bearings. Hence, db44 is adopted in this paper. As shown in Fig. 4, Samples are firstly decomposed into forty coefficients at three depths, and then the kurtosis and energy of the 8 last coefficients (third depth of decomposition) are calculated. As result another feature set containing 16 features is obtained.

Empirical Mode Decomposition EMD. Empirical mode decomposition (EMD) is relatively new method of signal processing which was applied in bearings and gears fault diagnosis of rotating machinery (Liu et al. 2005; Mahgoun et al. 2016). It does not use a priori determined basis functions and can iteratively decompose a complex signal

Table 1. List of the extracted features

No	Feature	No	Feature
1	Mean	18	Kurtosis of coefficient 3.3
2	RMS	19	Kurtosis of coefficient 3.4
3	Skewness	20	Kurtosis of coefficient 3.5
4	Kurtosis	21	Kurtosis of coefficient 3.6
5	Peak factor	22	Kurtosis of coefficient 3.7
6	Peak to peak value	23	Energy of coefficient 3.0
7	Clearance factor	24	Energy of coefficient 3.1
8	Shape factor	25	Energy of coefficient 3.2
9	Impulse factor	26	Energy of coefficient 3.3
10	PSD in [895–1321 Hz]	27	Energy of coefficient 3.4
11	PSD in [2004–2430 Hz]	28	Energy of coefficient 3.5
12	PSD in [3113–3539 Hz]	29	Energy of coefficient 3.6
13	PSD in [4222–4648 Hz]	30	Energy of coefficient 3.7
14	PSD in [5331–5757 Hz]	31	Kurtosis of the 1st IMF
15	Kurtosis of coefficient 3.0	32	Kurtosis of the 2nd IMF
16	Kurtosis du coefficient 3.1	33	Kurtosis of the 3rd IMF
17	Kurtosis of coefficient 3.2		

into a finite number of zero mean oscillations named intrinsic mode functions (IMFs). Each resulting elementary component (IMF) can represent the local characteristic of the signal (Mahgoun et al. 2016).

Samples were decomposed into a number of IMFs, and then the kurtosis of the three first IMFs is calculated.

The feature extraction operation was repeated with all samples of the three operating modes (normal, with defect level 1, and defect level 2). Table 1 summarize the list of the extracted features.

4 Feature Selection

From the above section, one can understand that there will be thirty three (33) features extracted for classification of samples belonging to three different classes. However, the entire feature set will not be used for the classification. Some of the features contain redundant information which may unnecessarily increase the complexity. This problem is frequently found in almost all pattern recognition problems. The challenge is to find out the most pertinent features and eliminate the redundant features to increase the classification accuracy.

In this study, we propose a filter based feature selection method. First, features are ranked in decreasing order based on their evaluation with a selection criterion. Afterword Pareto method is used to select the optimal feature subset according to features evaluations, then the corresponding classification accuracies using SVMs are tabulated. Three different criterions are compared: Fisher criterion, correlation criterion, and Signal to noise ratio (SNR).

4.1 Pareto Based Feature Selection Method

Pareto is a technique used for decision making based on the Pareto Principle, known as the 80/20 rule (Kramp et al. 2016). It is a decision-making technique that statistically separates a limited number of input factors as having the greatest impact on an outcome, either desirable or undesirable. Pareto analysis is based on the idea that 80% of a project's benefit can be achieved by doing 20% of the work. This ratio is used in this study to select the optimal feature subset from the initial set. The selected features are those cumulating 80% of the selection criterion score. This can be realised as follow:

1. The first step is to evaluate the score of each feature using a selection criterion,
2. The second step is to rank features in decreasing order according to their scores,
3. Compute the cumulative percentage of each feature,
4. Plot a curve with features on x- and cumulative percentage on y-axis,
5. Plot a bar graph with features on x- and percent frequency on y-axis,
6. Draw a horizontal dotted line at 80% from the y-axis to intersect the curve. Then draw a vertical dotted line from the point of intersection to the x-axis. The vertical dotted line separates the important features (on the left) and trivial features (on the right).

4.2 Selection Criterions

In the proposed method, features are selected according to their evaluation using three different criterions: fisher score, correlation criterion, and Signal to noise criterion. Also, the effect of these criterions on classification accuracy will be discussed in Sect. 5.

Fisher score. The idea is that features with high quality should assign similar values to instances in the same class and different values to instances from different classes. With this intuition, the score for the *i-th* feature *S(i)* will be calculated by Fisher Score as (Duda et al. 2000):

$$s(i) = \frac{\sum_{j=1}^{c} n_j \left(\bar{\mu}_{ij} - \bar{\mu}_i\right)^2}{\sum_{j=1}^{c} n_j \rho_{ij}^2} \tag{1}$$

where $\bar{\mu}_{ij}$ and ρ_{ij} are the mean and the variance of the i-th feature in the j-th class respectively, *nj* is the number of instances in the j-th class, and $\bar{\mu}_i$ is the mean of the i-th feature, c is the number of classes.

Correlation Criterion. The Correlation criterion evaluates features on the basis of the hypothesis that good feature is highly correlated with the classification. This correlation is measured using "Bravais-Pearson" criterion given by the following equation (Dash and Liu 2003):

$$C(i) = \frac{\sum_{k=1}^{m} (\mu_{ik} - \bar{\mu}_i)(y_k - \bar{y})}{\sqrt{\sum_{k=1}^{m} (\mu_{ik} - \bar{\mu}_i)^2 \sum_{k=1}^{m} (y_k - \bar{y})^2}} \tag{2}$$

where $\bar{\mu}_i$ et $\bar{y}$ are the mean of the *i-th* feature and class labels of data respectively. m is the number of all instances.

Signal to Noise Ratio (SNR). The signal to noise ratio (SNR) identifies the expression patterns with a maximal difference in mean expression between two classes and minimal variation of expression within each class (Mishra and Sahu 2011).

$$SNR(i) = \frac{|\bar{\mu}_{i1} - \bar{\mu}_{i2}|}{(\sigma_{i1} - \sigma_{i2})} \tag{3}$$

Where $\bar{\mu}_{i1}$ and $\bar{\mu}_{i2}$ denote the mean of the i-th feature in class 1 and class 2 respectively. $\sigma 1$ and $\sigma 2$ are the standard deviations for the i-th feature in each class.

5 Results and Discussion

From the initial feature set, the best features have been selected using Pareto-Fisher based feature selection algorithm, Pareto-correlation based feature selection algorithm, and Pareto-SNR based feature selection algorithm given in the above section.

Data composed of 240 samples was divided into two equally subsets. The first one is used for training SVMs, while the second is used for the test. SVMs accuracy is evaluated by the number of misclassified samples in the test. Based on results of previous work (Ziani et al. 2017), SVMs is trained with an RBF kernel and OAO strategy for multiclass SVM is adopted.

In the first time the SVMs classifier has been trained with the initial feature set composed of 33 features, then it has been trained with the optimal feature subset selected with different algorithms and the results are tabulated as follows:

Figure 5 shows the Pareto curve in the case of features selection using Fisher score. Features are ranked according to their scores, and then the selected features are those cumulating 80% of Fisher criterion. In this case, the optimal feature subset is composed of the following features: 1, 10, 6, 30, 2, 8, and 27.

From Tables 2 and 3, one can understand that three algorithms have selected the pertinent feature subset in different manner. However, looking at a problem in classification accuracy view point, it is clear that the classification accuracy was improved with the selected features in all cases. The Pareto-fisher gives 100% with only seven features, Pareto-correlation gives also 100% but with ten features, and finally Pareto-SNR gives 97.5% with 13 features.

Figure 6 shows 3D scatter plot of data with the entire feature set. This plot is performed using Principal Components Analysis (PCA) where data is projected on three Principal Components: PC1, PC2, and PC3. It is important to note that PCA is used here for data visualization but not for selection purpose. Figures 7, 8 and 9 show plots of data with pertinent features selected using the three criterions. It is clear that data is well separated using the selected features which explain the improvement of classification accuracy. The best data separation is obtained using the selected features by Pareto-fisher algorithm.

From Table 2, one can understand that the selected features are not the same in the three cases. This is logical since different criterions were used. However some features are selected by the three algorithms which confirm their discriminant ability. These features are: the mean, peak to peak, PSD calculated in the band [895–1321 Hz], and Energy of coefficient 3.7. The mean and Peak to peak values quantify the level of vibration. When any fault occurs in a gear, the level of vibration increase and the values of these features increase consequently. This can be confirmed in Fig. 3 where the level of vibration increases with the level of defect. PSD is a measure of the power of signal in frequency domain. When fault appear, PSD calculated around the meshing frequency increase significantly. This can be explained by the modulation phenomena characterized by the production of sidebands around the meshing frequency.

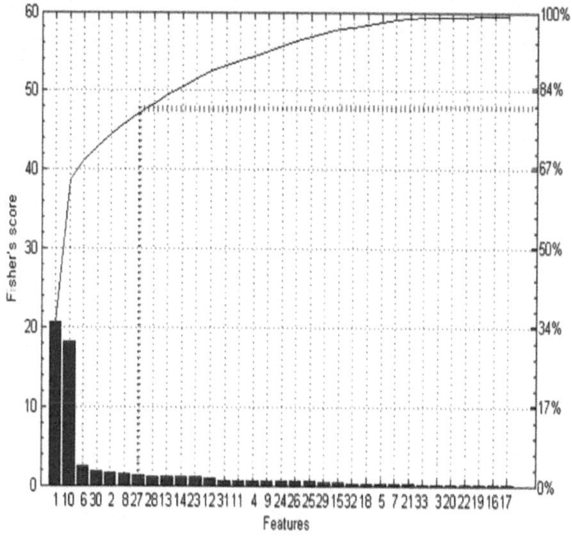

Fig. 5. Pareto curve with Fisher score

Table 2. Optimal features subsets

Selection method	Optimal feature subset
Pareto-Fisher	1, 10, 6, 30, 2, 8, 27
Pareto-correlation	10, 30, 4, 13, 15, 25, 27, 29, 18, 1
Pareto-SNR	1, 10, 11, 6, 2, 23, 28, 14, 12, 8, 24, 13, 26

Table 3. SVM classification accuracy

Inputs	Number of features	SVM accuracy (%)
The entire feature set	33	95.83
Features selected with Pareto-fisher	7	100
Features Selected with Pareto-correlation	10	100
Features Selected with Pareto-SNR	13	97.5

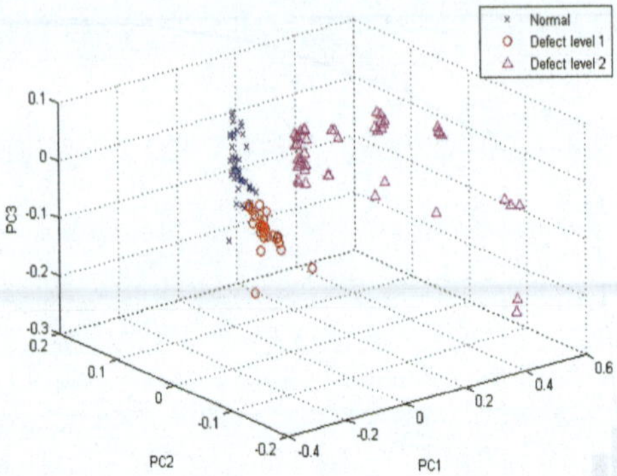

Fig. 6. 3D scatter plot of data with the entire feature set (33 features)

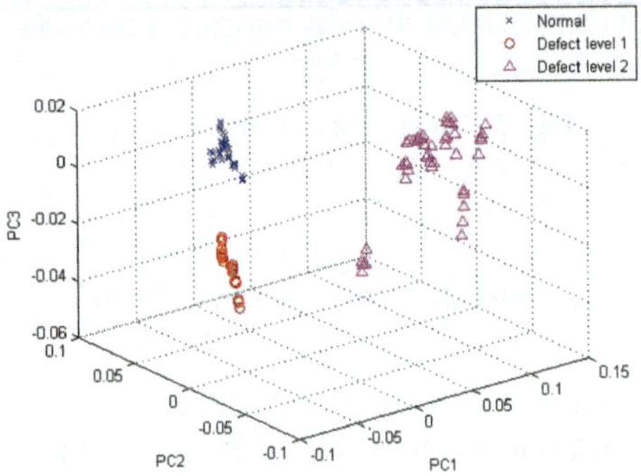

Fig. 7. 3D scatter plot of data with features selected by Pareto-Fisher algorithm (7 features)

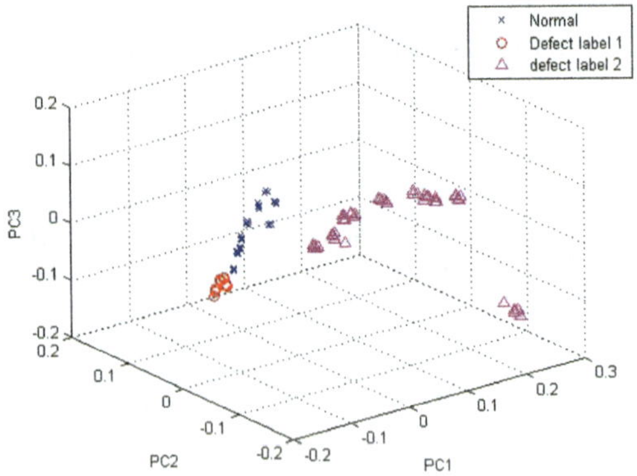

Fig. 8. 3D scatter plot of data with features selected by Pareto-correlation algorithm (10 features)

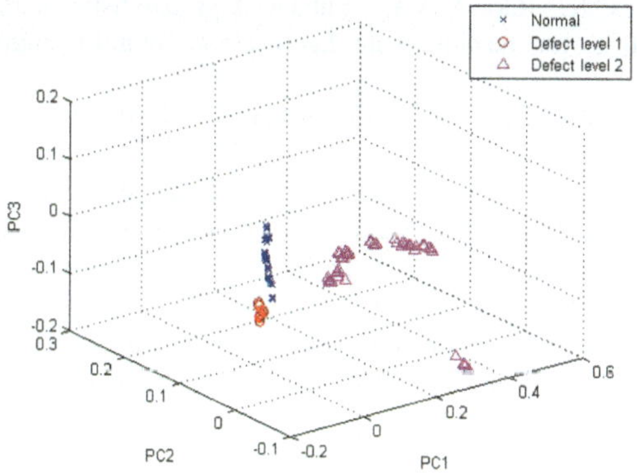

Fig. 9. 3D scatter plot of data with features selected by Pareto-SNR algorithm (13 features)

6 Conclusion

In this paper, an investigation has been made on different feature selection criterions and their effect on classification also studied. Different features were extracted from the vibration data using different signal processing methods. There were totally thirty three features out of which certain features may not be use for classification. The optimal feature subset was selected according three different criterions such as: Fisher score, correlation criterion, and Signal to Noise criterion. Their results and corresponding classification accuracies have been tabulated. Pareto method has been used to define the number of features to be selected. It can be concluded that Pareto-fisher based feature selection algorithm with SVMs classifier seem to perform better for this application. However, other algorithms also may suit for some other applications. Our future work will focus on a more comprehensive fault diagnosis of rotating machinery based on the unsupervised learning methods.

References

Abdul Rahman, A.G., Chao, O.Z., Ismail, Z.: Effectiveness of impact-synchronous time averaging in determination of dynamic characteristics of a rotor dynamic system. Measurement **44**, 34–45 (2011)

Bartkowiak, A., Zimroz, R.: Dimensionality reduction via variables selection - linear and nonlinear approaches with application to vibration-based condition monitoring of planetary gearbox. Appl. Accoustics **77**, 169–177 (2014)

Baydar, N., Ball, A.: A comparative study of acoustic signals in detection of gear failures using Wigner-Ville distribution. Mech. Syst. Signal Process. **15**, 1091–1107 (2001)

Burges, C.A.: Tutorial on support vector machines for pattern recognition. Data Min. Knowl. Disc. **2**, 955–974 (1998)

Cameron, B.G.: Final report on CH-46 Aft transmission seeded fault testing, Research Paper RP907. Westland Helicopters Ltd, UK (1993)

Chang, R.K.Y., Loo, C.K., Rao, M.V.C.: Enhanced probabilistic neural network with data imputation capabilities for machine-fault classification. Neural Comput. Appl. **18**, 791–800 (2009)

Dash, M., Liu, H.: Consistency-based search in feature selection. Artif. Intell. **151**, 155–176 (2003)

Duda, R., Hart, P., Stork, D.: Pattern Classification, 2nd edn. Wiley, Hoboken (2000)

Goyal, D., Vanraj, B.S., Pabla, S., Dhami, S.: Condition monitoring parameters for fault diagnosis of fixed axis gearbox: a review. Arch. Computat. Methods Eng. **24**(3), 543–556 (2016). https://doi.org/10.1007/s11831-016-9176-1

Gryllias, K.C., Antoniadis, I.A.: A support vector machine approach based on physical model training for rolling element bearing fault detection in industrial environments. Eng. Appl. Artif. Intell. **25**, 326–344 (2012)

Konar, P., Chattopadhyay, P.: Bearing fault detection of induction motor using wavelet and support vector machines (SVMs). Appl. Soft Comput. **11**, 4203–4211 (2011)

Kotsiantis, S.B.: Feature selection for machine learning classification problems: a recent overview. Artif. Intell. Rev. **42**(1), 157 (2011). https://doi.org/10.1007/s10462-011-9230-1

Kramp, K.H., Van Det, M.J., Veeger, N.J.G.M.: The Pareto analysis for establishing content criteria in surgical training. J. Surg. Educ. **73**, 892–901 (2016)

Kudo, M., Sklansky, J.: Comparison of algorithms that select features for pattern classifiers. Pattern Recogn. **33**, 25–41 (2000)

Liu, B., Riemenschneider, S., Xub, Y.: Gearbox fault diagnosis using empirical mode decomposition and Hilbert spectrum. Mech. Syst. Signal Process. **17**, 1–17 (2005)

Loughlin, P., Cakrak, F.: Conditional moments analysis of transients with application to the helicopter fault data». Mech. Syst. Signal Process. **14**, 515–522 (2000)

Mahgoun, H., Chaari, F., Felkaoui, A., Haddar, M.: Early detection of gear faults in variable load and local defect size using ensemble empirical mode decomposition (EEMD). In: Advances in Acoustic and Vibration, Proceeding of the International Conference on Acoustic and Vibration (ICAV 2016), Hammamet, Tunisia (2016)

Mishra, D., Sahu, B.: Feature selection for cancer classification: a signal-to-noise ratio approach. Int. J. Sci. Eng. Res. **2**, 1–7 (2011)

Nandi, A.K., Liu, C., Wong, M.L.D.: Intelligent vibration signal processing for condition monitoring. In: International Conference Surveillance 7, Institute of Technology of Chartres, France, 29–30 October 2013

Rafiee, J., Arvani, F., Harifi, A., Sadeghi, M.-H.: Intelligent condition monitoring of a gearbox using artificial neural network. Mech. Syst. Signal Process. **21**, 1746–1754 (2007)

Rafiee, J., Rafiee, M.A., Tse, P.W.: Application of mother wavelet functions for automatic gear and bearing fault diagnosis. Expert Syst. Appl. **37**, 4568–4579 (2010)

Scholkopf, B.: SVMs-a practical consequence of learning theory. IEEE Intell. Syst. **13**, 18–19 (1998)

Tang, J., Alelyani, S., Liu, H.: Feature selection for classification: a review. In: Data Classification: Algorithms and Applications, p. 37. CRC Press (2014)

Vapnik, V.N.: Statistical Learning Theory. Wiley, New York (1998)

Williams, W.J., Zalubas, E.J.: Helicopter transmission fault detection via time-frequency, scale and spectral methods. Mech. Syst. Sig. Process. **14**, 545–559 (2000)

Worden, K., Staszewski, W.J., Hensman, J.J.: Natural computing for mechanical systems research: a tutorial overview. Mech. Syst. Sig. Process. **25**, 4–111 (2011)

Yang, B.S., Hwang, W.W., Han, T.: Fault diagnosis of rotating machinery based on multi-class support vector machines. J. Mech. Sci. Technol. **19**, 846–859 (2005)

Zhang, Z., Wang, Y., Wang, K.: Fault diagnosis and prognosis using wavelet packet decomposition, Fourier transform and artificial neural network. J. Intell. Manuf. **24**, 1213–1227 (2013)

Ziani, R., Felkaoui, A., Zegadi, R.: Bearing fault diagnosis using multiclass support vector machines with binary particle swarm optimization and regularized Fisher's criterion. J. Intell. Manuf. **28**(2), 405–417 (2017)

Intelligent Gear Fault Diagnosis in Normal and Non-stationary Conditions Based on Instantaneous Angular Speed, Differential Evolution and Multi-class Support Vector Machine

Semchedine Fedala[1(✉)], Didier Rémond[2], Ahmed Felkaoui[1], and Houssem Selmani[1]

[1] LMPA, Applied Precision Mechanics Laboratory, Institute of Optics and Precision Mechanics, Setif-1-University, 19000 Setif, Algeria
semchedinef@yahoo.fr, a_felkaoui@yahoo.fr, selmani.houssem@gmail.com

[2] LaMCoS, UMR5259, CNRS, INSA, Lyon 69621, Villeurbanne CEDEX, France
didier.remond@insa-lyon.fr

Abstract. The gearboxes are among the most important elements of rotating machines and consequently they require an effective condition monitoring strategy. However, many machines operate over a wide range of the rotational speed and most analysis of rotating machines are based on investigating the vibrations with a constant speed. Therefore, techniques developed for constant conditions cannot be applied directly.

The angularly sampled Instantaneous Angular Speed (IAS) carry a considerable amount of information on the health and usage status of rotating machinery. Thus, it represents a potential source of relevant information in intelligent fault detection and diagnosis systems, but also to construct Feature Vector (FV) to further get robust and effective classification methods for different running speed or load conditions.

This paper presents an intelligent gear fault diagnosis based on Instantaneous Angular Speed (IAS), Differential Evolution (DE) and multi-class Support Vector Machine (SVM) in normal and non-stationary conditions. For this purpose, features are extracted from IAS. Then, the DE selection algorithm is applied in order to select the most relevant features. The classification is performed by SVM in order to improve the detection and identification of gear defects. The methodology is applied in normal and non-stationary conditions, with six pinion fault conditions. The experimental results prove that the proposed method is able to detect the fault conditions of the gearbox effectively.

Keywords: Gearbox fault diagnosis · Non-stationary conditions
Instantaneous Angular Speed · Differential Evolution
Support Vector Machines

A. Felkaoui et al. (Eds.): SIGPROMD'2017, ACM 12, pp. 16–33, 2019.
https://doi.org/10.1007/978-3-319-96181-1_2

1 Introduction

Condition monitoring of rotating machines is one of the areas of engineering that is gaining importance in industry. Its role is to ensure the operation continuity of mechanical systems in factories, in order to limit production losses due to unexpected failures (Randall 2011; Boulenger and Pachaud 2009). This monitoring can be automated by implementing classification methods (Dubuisson 1990). The performances of these methods are closely related to the relevance of fault indicators making up the feature vectors (FV) of these classification methods. The FV must be able to describe the different operation modes or system damage, and also reflect the precise definition of the classes that represent the different operation modes (Vachtsevanos et al. 2006). Generally, the indicators are based on signals analysis provided by the sensors installed on the monitored system (accelerations, speeds, torques, currents, voltage … etc.) and must be also constructed automatically to ensure the most robust analysis. Current research on the automation of vibration diagnosis is mainly based on indicators extracted from the Time Sampled Acceleration signals (TA). The major drawback of these signals is their sensitivity to operating speed conditions, particularly in non-stationary conditions. Therefore, there is variation in the number of samples acquired by revolution but also changes in excitation frequencies related to the discrete geometry rotation. In this context, it is difficult (or impossible in non-stationary conditions) to identify a characteristic frequency in the spectrum in an automated manner. One alternative is to have angularly sampled signals, which ensures a constant integer number of samples per revolution and by getting rid of speed fluctuations. Furthermore, the assessment of the interest frequency component level may be biased by the phenomenon of "picket fence effect" (Rémond 1998). A detailed description of the solutions to obtain the angularly sampled signals such as the Angularly sampled Acceleration (AA), the Transmission Error (TE) and the Instantaneous Angular Speed (IAS) can be found in the literature (Renaudin et al. 2010; Kong 1987; Li et al. 2005; Fyfe and Munck 1997). Specially, it has been shown that the IAS provides sensitive and robust indicators for the early detection of defects (Renaudin et al. 2010). Indeed, the major interest of using IAS signal analysis is that any defect or failure present in a rotating machine will inevitably change the rotating dynamics of the machine, and as a result, the instantaneous angular speed of the shaft will theoretically vary. Hence, analysis of IAS signals will provide valuable fault-related information on the machine. Most recently, the IAS has widely been used in the monitoring of gear transmissions (Rémond 1998; Leclère and Hamzaoui 2014), bearings (Renaudin et al. 2010), machining (Lamraoui et al. 2014) and intelligent diagnosis of gear faults (Fedala et al. 2015, 2016).

On the other hand, from the intelligent fault diagnosis point of view, some features extracted from IAS or from other signals are redundant or irrelevant (Vachtsevanos et al. 2006). So, if the whole feature set is employed by the classification method directly, it may lead to misclassification and thus misdiagnosis. Hence, to improve the diagnosis accuracy, a set of significant features that discriminate between the observations in different failure modes need to be selected from the original feature set (Fedala et al. 2009; Samanta et al. 2004). To work around this problem, selection methods were applied in diagnostics according to the choice and experience accumulated by researchers (Kudo and Sklansky 2000; Khushaba et al. 2011). These methods can be divided into two categories: filters and wrappers. Filter based feature selection methods are generally faster because they depend on some type of estimation of the importance of individual features or subset of features. Whereas wrapper-based methods are more accurate, because the quality of the selected subset of features is evaluated using a learning method. There are many optimization methods that are used to find the optimal number of features, but that differ in their optimality and computational cost. Such methods add the important features, one by one at each iteration, until the best performance is reached. Among these methods: Genetic methods (GA) (Haupt and Haupt 2004), Ant Colony Optimization (ACO) (Dorigo and Stützle 2004), Particle Swarm Optimization (PSO) (Kennedy et al. 2001) and Differential Evolution optimization method (DE) (Price et al. 2005). In this work, a modified DE optimization method proposed in Khushaba et al. (2011) and abbreviated as DEFS technique is used for feature subset selection. The advantage of this method is that the practical results indicate the significance of the proposed method in comparison with other feature selection methods. Another reason for choosing DEFS is that it selects the relevant features automatically and not perturb or hide their physical meaning, which is very important for interpretability of the classification and diagnosis results.

In this paper, IAS is determined to monitor different operating modes. For this purpose, features are extracted from angular and frequency domains. Then, the DEFS is applied in order to select the most relevant features. The classification is performed by multi-class Support Vector Machines (SVM) for the improvement of the detection and identification of gear defects. The methodology is applied in healthy conditions, then for five pinion faults with different running speed and load conditions.

The main contribution of this paper is the use of new relevant indicators even in the non-stationary conditions, to feed the classification methods, which requires the analysis of the IAS signals estimated from the signals delivered by the optical encoders. Knowing that the current trend of bearing manufacturers and rotary machines is to integrate angular encoders in their products in order to obtain different types of controls (speed control, position…). The use of these signals makes it possible to carry out acquisitions directly in angular sampling without any additional constraints.

The paper is structured as follows. The first part provides an overview of IAS measuring principle. Then, the description of the experimental device as well as the various test conditions. Afterwards, we present the analysis of the characteristics of the measured variables and the different indicators introduced as a Feature Vector VF. After, we show the selection procedure using DEFS. Finally, we provide the SVM classification results and related discussion in order to show the advantages of the proposed approach and give the final conclusion.

2 Measuring Principle

The use of high resolution optical encoders, that is to say having a large number of pulses per revolution, offers the possibility to measure the IAS. Several methods are used to estimate IAS from a pulse signal. Generally, these methods can be categorized into two groups. The first technique, called (ADC)-based methods (Fyfe and Munck 1997), is based on the use of a standard analog to digital converter with an anti-aliasing filter to acquire the angle encoder signal, using a lower sampling rate (typically several tens of kHz), and to determine event's times of the encoder by numerical processing like upsampling or interpolation. The second technique, known as timer/counter-based methods (Kong 1987; Li et al. 2005), illustrated in Fig. 1, consists to use a high frequency counting approach: a high frequency pulse signal is used as reference, typically several tens of MHz, and an electronic device is used to count the number of pulses of the high frequency clock between two events of the angle encoder signal. This method will be used in this paper.

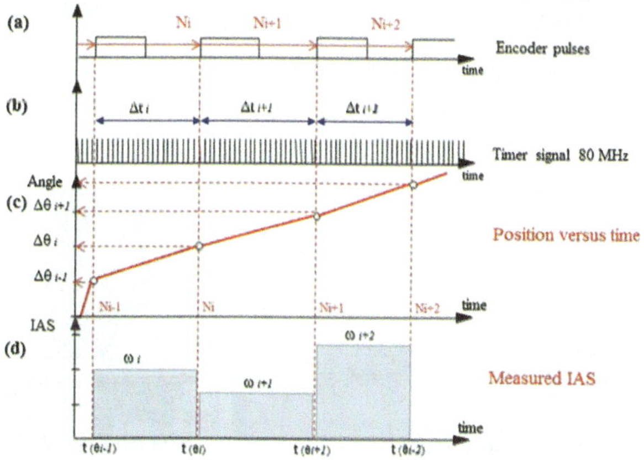

Fig. 1. Principle of instantaneous angular speed measurement with the counting method: (a) encoder pulses, (b) high frequency clock pulses, (c) angular position for the shaft and (d) IAS estimation

The reconstruction of the IAS signal is directly estimated by

$$\omega_i = \frac{2\pi f_h}{N_f\, n_i} \ \text{(rad/s)} \tag{1}$$

where f_h is the clock frequency of the counter board; n_i the number of pulses/counts for each angle interval i between two rising edges of the encoder signal; N_f is the resolution of the optical encoder. The index used i on ωi denotes instantaneous.

3 Test Bench and Experimental Protocol

The test bed Fig. 3a used in this study consists of two rotating shafts, on which are mounted a pinion and a spur gear offering a gear ratio of 25/56 respectively. To compare the effectiveness of the analysis methods, we used six pinions, the first one is referred as Good (G), whereas the others have several different types of defects: a Root

Crack (RC), a Chipped Tooth in Width (CTW), a Chipped Tooth in Length (CTL), a Missing Tooth (MT) and General Surface Wear (GSW) as shown in Fig. 2. Three pinions are simultaneously mounted on the input shaft of the gearbox; the engagement change is done by a simple axial movement of the wheel on its axis Fig. 3b.

Fig. 2. View of six used pinions. (a) G. (b) RC. (c) CTW. (d) CTL. (e) MT. (f) GSW

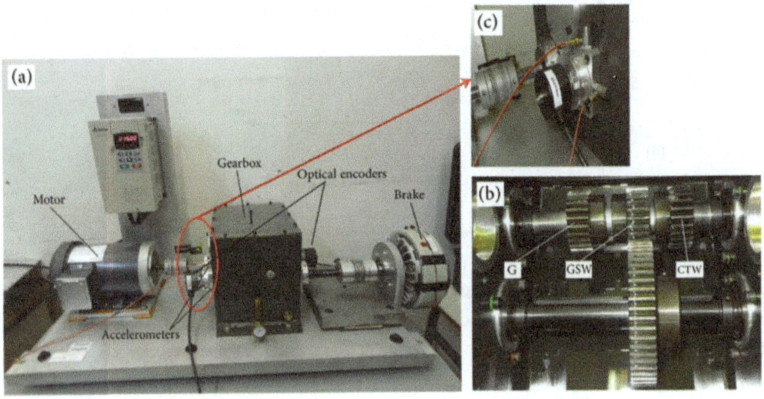

Fig. 3. The test bed (a), location of pinions (b) and optical encoder (c)

The input shaft is driven by an electric motor controlled in rotational speed. The engine ensures a maximum speed of 3600 rpm. The output shaft is connected to a magnetic powder brake capable of generating different resistive torques. To measure the angular positions of the shafts, two optical encoders of 2500 pulses per revolution are mounted at the free ends of the two shafts of the gearbox Fig. 3c. The clock frequency of the counting acquisition system is 80 MHz, generally considered sufficient to locate the rising edges of the encoder signals. The encoders signals are recorded during 1200 revolutions of the input shaft. The angular positions have been measured for different operating conditions by varying the rotation speed and the resistant torque for each of the six gears used (Table 1). Each test is repeated ten times for normal conditions (Fig. 5a) and five times for non-stationary conditions (Fig. 5b and c), in order to have a sufficient number of signals for the training and testing of SVM. In total, 1260 records have therefore been made, 210 records for each class of operation.

For the non-stationary running conditions two strategies have been used:

- Load variation: for the five RPMs used (Table 1) the load balances suddenly from the no-load to under load operation (Fig. 5b), repeated several times during the acquisition time. Two resistive torques are used, 5 and 8 Nm.
- Speed variation: for the two couples used (5 and 8 Nm), speed increases gradually until an approximated value of 50 Hz, then decreases with the same manner till the end of acquisition (Fig. 5c).

Table 1. Running conditions

Fault description	Denomination	Conditions		RPMs (r/min)	Load (N m)	Number of signals
Good	G	Normal (stationary) (S)		900, 1200, 1500, 1800, 2400	5, 8, 11	900
Root Crack	RC					
Chipped Tooth in Width	CTW	Non-stationary (NS)	Load (NSL)	900, 1200, 1500, 1800, 2400	Load variation – 0 and 5 – 0 and 8	300
Chipped Tooth in Length	CTL					
Missing Tooth	MT		Speed (NSS)	Speed variation	5, 8	60
General Surface Wear	GSW					

4 Experimental Part

The flowchart in Fig. 4 shows a complete overview of the techniques used in this study. From records made on the test bench, IAS signal is employed in order to extract different types of indicators to build a FV, which afterwards, will be used, with and without selection by DEFS algorithm, to the training and the testing of the SVM classifier.

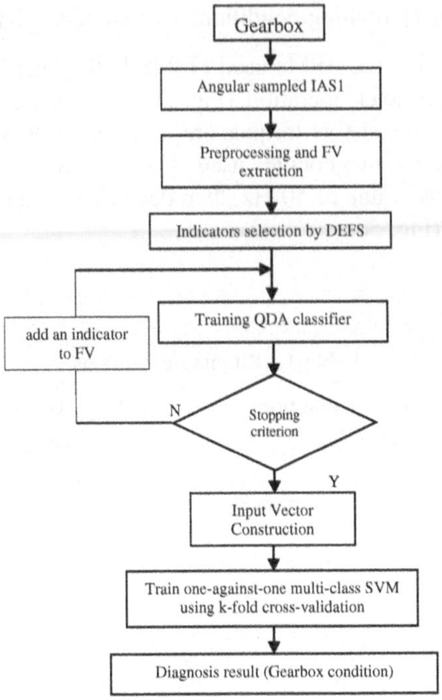

Fig. 4. Flowchart of the SVM based fault diagnosis system

4.1 Feature Extraction

The feature extraction stage is the most important part in the pattern recognition process. The purpose of feature extraction is twofold; firstly, feature extraction is an attempt to reduce the dimensionality of the data presented to the classifier, without diminishing the content presented in the data. Secondly, feature extraction is intended to turn signals into indicators information that the classifier can use more usefully (Dubuisson 1990).

4.1.1 Signal Analysis (Angular Features Extraction)

The IAS1 signals are calculated from (1), using the encoder signal mounted on the input shaft of the gearbox where pinions with defects are mounted on. Although Fig. 5 a and b clearly show the presence of a defect that is manifested by an increase in IAS1. However, only the IAS fluctuations will make it possible to determine the state of the pinion and not the change in the speed of rotation which is likely to give inaccurate values, in particular scalar indicators dedicated to the monitoring of defects. As a result, a normalization of these signals is indispensable. Considering the almost constant speed during a lap, this normalization is done by subtracting the average of the speed of each lap from the points of the IAS1 constituting the same lap.

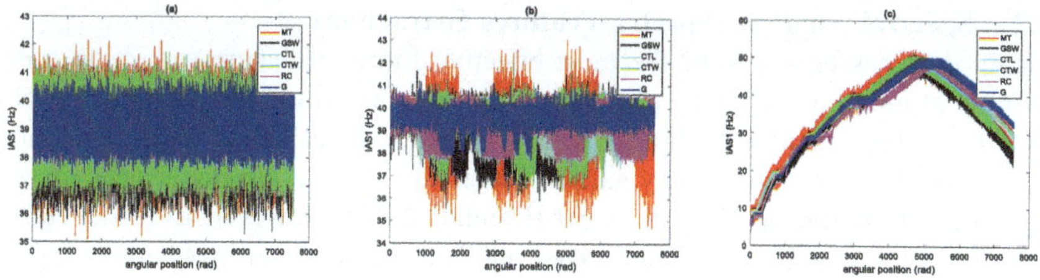

Fig. 5. Presentation of IAS1 signals in normal (a) and non-stationary conditions: (b) Torque variation and (c) speed variation, for pinions: good (G), Root Crack (RC), Chipped Tooth in Width (CTW), Chipped Tooth in Length (CTL), Missing Tooth (MT) and General Surface Wear (GSW)

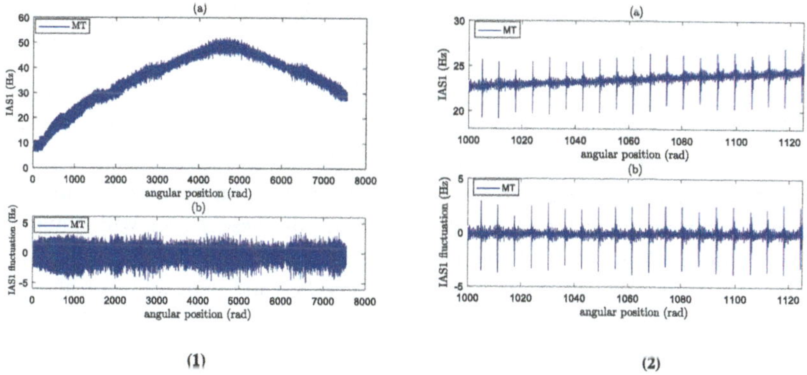

Fig. 6. Presentation of IAS1 signal in non-stationary conditions (speed variation) for pinion with MT defect. (1) IAS signal (a) before and (b) after normalization. (2) zoomed view of IAS signal (a) before and (b) after normalization

Figure 6a and b show respectively the IAS1 signals, recorded for the MT defect, before and after normalization. Figure 6a and b show the zoomed view of these signals for 20 shaft revolutions. As can be seen, even if the two signals have periodic pulses, the velocity fluctuations in the case of the normalized IAS oscillate around the zero and not around a variable value. Therefore, the advantage of normalization is related to the value of scalar indicators such as kurtosis and RMS, which had respective values of 2.99 and 37.61 before and which regain after normalization a respective values of 8.62 and 0.43 which are more representative of the pulses generated by the defect. Thus, the IAS1 signals are processed to extract five angular domain features commonly used in literature, which are RMS, variance, crest factor, kurtosis and skewness. The definitions of these features can be found in (Mohanty 2015).

4.1.2 Spectral Analysis (Spectra Features Extraction)

Spectral data has been one of the most effective forms of feature extraction used in condition monitoring. As many of the machines monitored are rotational, many of the faults that exhibit themselves are frequency related, where the machines kinematic is also known, it is a comparatively simple matter to calculate the frequencies at which certain defects would be likely to occur (Randall 2011). However, to read frequency plots, identify harmonic peaks, and give confident diagnoses of problems is a skilled task, and requires experience. Spectral information is still very useful for providing information for classifiers, and as a result, it was used as one of the methods of features extraction.

The spectral field of the IAS1 signals presented in Fig. 7, give a considerable advantage, due to the observation frequency of gears being not changed by the rotational speed, but are directly observable at the main orders:

- channel 1 and its harmonics for localized defects,
- channel 25 corresponding to the pinion teeth number (Z = 25) and its harmonics, for generalized defects.

So, the presence of the fault on the pinion causes a number of events per revolution, the significant increase in the peak amplitude of the frequency channel corresponding either to the number of teeth of the pinion (Z = 25) for generalized defects, or on the frequency channel 1 for localized defects. We also remark an increase in energy of the intermediate levels. It is found that the positions of these peaks will remain fixed despite variations in speed from one test to another, whereas the amplitudes vary in a different way from one frequency channel to another and depending on the type of fault. Consequently, it is a source of building highly relevant indicators. These figures are used to track with precision the frequency components associated to the different types of defects and to the supervised geometrics (number of teeth of the pinion), whether they are localized or generalized. These amplitudes are subsequently used as indicators in the FVs.

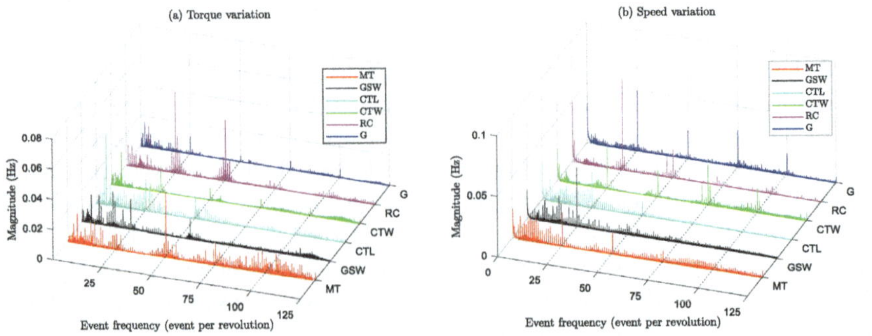

Fig. 7. Event spectra of IAS1 signals in non-stationary conditions (a) torque variation (b) speed variation, using pinions: good (G), Root Crack (RC), Chipped Tooth in Width (CTW), Chipped Tooth in Length (CTL), Missing Tooth (MT) and General Surface Wear (GSW)

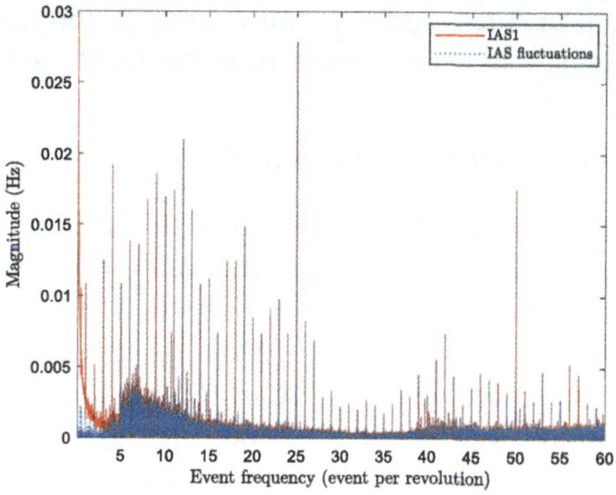

Fig. 8. Event spectra of IAS1 signals in non-stationary conditions (speed variation, torque 8 Nm) before and after normalisation for the pinion with MT defect

Moreover, the spectra of the IAS1 signals before and after normalization (Fig. 8) reveal that the amplitudes of the different components are perfectly superimposed and that normalization allows only to get rid of the phenomenon associated with the effect of speed increase in bass orders by eliminating the dominant harmonic (at 0 events/rev) which represents the mean value of the signal. Therefore, this normalization operation only improves the relevance of the scalar indicators.

4.2 Feature Vector

We propose to use a FV of 17 features according to the flowchart given in Fig. 4. They are summarized in Table 2. All used signals are processed to extract:

Table 2. Description of the features

Indicators	Domain	IAS1
1	Angular	RMS
2		Variance
3		Crest factor
4		Kurtosis
5		Skewness
6	Frequency	The level of order 1
7		The level of order 25
8		The level of order 50
9		The level of order 75
10		The level of order 100
11		The level of order 125
12		The sum of the levels of the 2nd to 24th order
13		The sum of the levels of the 26th to 49th order
14		The sum of the levels of the 51th to 74th order
15		The sum of the levels of the 76th to 99th order
16		The sum of the levels of the 101th to 124th order
17		The sum of the levels of the 1st to 125th order

- five angular domain features: RMS, variance, crest factor, kurtosis and skewness,
- twelve orders frequency domain features from the IAS1 signals.

4.3 Feature Selection by DEFS Algorithm

The preceding sections present the methodology to extract the features by signal processing. Some of these features could be irrelevant or redundant. The Differential Evolution (DE) algorithm is one of the powerful selection techniques. It selects a subset of features based on a simple optimization method that has parallel, direct search, good convergence and fast implementation properties (Price et al. 2005). The DEFS algorithm based on modifying the DE float-number optimizer in a combinatorial optimization problem like feature selection, is applied in this paper due to the superiority of obtained performances compared to all other methods (Khushaba et al. 2011). The readers are invited to consult this reference for more details about this algorithm.

We used this algorithm to find the most important features, it is applied to the complete set of features to select the most relevant, it consists of four basic steps which are subset generation, evaluation, stopping criterion and result validation by the selected features which minimize the classification error rate (i.e., the number of misclassified observations divided by the number of observations) of the classifier algorithm QDA (Quadratic Discriminant Analysis) (Hastie et al. 2009).

Table 3 shows the selected indicators in detection and identification stages for stationary (S), non-stationary conditions: Torque variation (NSL) and speed variation (NSS), and combined conditions (C). It can be seen that the number of selected indicator in both stages increases (relatively to the combined conditions) when the number of VFs increase (1260 VFs).

Table 3. Indicators ranking

Indicators ranking	Detection				Identification			
	S	NSL	NSS	C	S	NSL	NSS	C
1	5	6	9	1	1	2	7	1
2	6	8	15	2	6	6	12	2
3	7	11		3	7	8		3
4	12	12		5	8	12		5
5	14	14		6	10			7
6	16			9				8
7				12				9
8				13				10
9				14				12
10				15				13
11				16				14
12								17

4.4 Classification Procedure

After this learning step, the classification of the tested experiment is performed by using multi-class SVM. Firstly, the detection is performed by testing if the FV belongs to the default class or to the healthy class. In case of default detection, the FV is compared with all the default classes for identification.

4.4.1 Support Vector Machine Theory

The support vector machine (SVM), proposed by Vapnik (1998), is recognized as the most powerful algorithms in classification (Widodo and Yang 2007; Bordoloi and Tiwari 2014; Samanta 2004). The basic principle of SVM is to separate two classes with optimal hyperplane which maximizes the margin between the separating hyperplane Fig. 9.

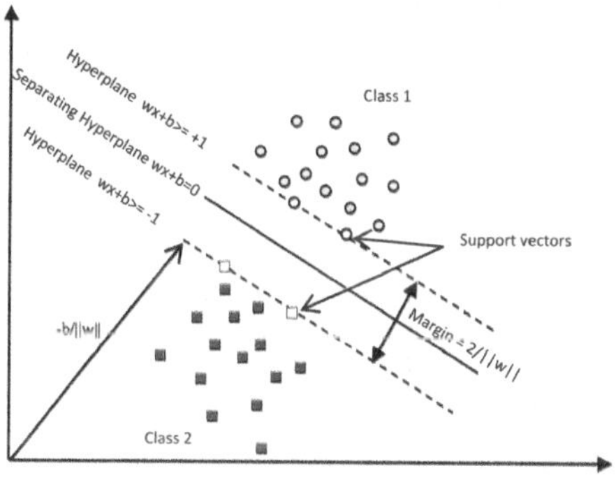

Fig. 9. Separation of two classes by SVM

To describe the algorithm of SVM, let us consider the set P that trains the SVM classifier:

$$P = (x_i, y_i), x_i \in R^m, y_i \in \{-1, 1\}_{i-1}^{n} i = 1, 2, \ldots, n \tag{2}$$

Where x_i represents an input vector containing m indicators of a n training set samples, while y_i is the desired output ($y_i = 1$ for positive class and $y_i = -1$ for negative class).

In the case of linearly separated data, the separating hyperplane $f(x) = 0$ can be expressed as:

$$f(x) = w^T x + b = \sum_{i=1}^{n} w_i x_i + b = 0 \tag{3}$$

Where w is a weight vector and the scalar $\boldsymbol{b}$ is the bias.

The separating hyperplane must satisfy the equation,

$$y_i f(x_i) = y_i(w^T x_i + b) - 1 \geq 1 \tag{4}$$

The Euclidean distance of any point that lies on either of the two hyperplanes is equal to $1/\|w\|$. Maximizing the margin $2/\|w\|$ is equivalent to minimizing $\|w\|^2$. The solution is found after resolving the following quadratic optimization problem:

$$\textit{mimimize} \quad \frac{1}{2}\|w\|^2 + C\sum_{i=1}^{n} \xi_i \tag{5}$$

$$\textit{subject to} \begin{cases} y_i(w^T x_i + b) \geq 1 - \xi_i, \\ \xi_i \geq 0, \end{cases} \quad i = 1, \ldots, n$$

Where C is the regularization parameter and ξ is the slack variables.
Using the Lagrangian optimization method, the above equation can be presented as:

$$\textit{maximize} \quad W(\alpha) = \sum_{i=1}^{n} \alpha_i - \frac{1}{2}\sum_{i,j=1}^{n} \alpha_i\alpha_j y_i y_j (x_i, x_j) \tag{6}$$

$$\textit{subject to} \quad \begin{cases} 0 \leq \alpha_i \leq C \\ \sum_{i=1}^{n} \alpha_i y_i = 0 \end{cases} \quad i = 1, \ldots, N$$

For the case of nonlinear separability in feature space, the kernel function is introducing in the last equation in order to transform the input vectors in to a high dimensional feature space, where the linear separation is possible. Thus, the inner product *(xi, xj)* (Eq. 6) is replaced by a kernel function *K(xi, xj)*, as shown in the following equation:

$$W(\alpha) = \sum_{i=1}^{n} \alpha_i - \frac{1}{2}\sum_{i,j=1}^{n} \alpha_i\alpha_j y_i y_j K(x_i, x_j) \tag{7}$$

Finally, based on the optimal hyperplane, the optimal classification function can be given as:

$$f(x) = \textit{sign}\left(\sum_{i,j=1}^{n} \alpha_i y_i K(x_i, x_j) + b\right) \tag{8}$$

The kernel functions commonly used in SVM's formulations are: linear, polynomial, sigmoid and radial basis function (RBF), etc. In this study, we opted for a *cubic* polynomial *kernel*. This function permits to separate perfectly learning samples and does not require any adjustment.

4.4.2 Multiclass SVM

The discussion above deals with binary classification where the class labels can take only two values: 1 and −1. Generally, in the rotating machineries there are several fault classes such as gear faults, mechanical unbalances, misalignments, bearing faults, etc. In the gear fault also several faults appear like the wear of teeth, the missing tooth, the chipped tooth, the root crack, etc. Consequently, an appropriate multi-class method is needed. A number of possible methods for this purpose are as follows (Chapelle et al. 1999):

- Modifying the design of the SVM to incorporate the multi-class learning directly in the quadratic solving algorithm,
- Combining several binary classifiers with two methods:
  - One-against-one (OAO), which applies pair comparisons between classes.
  - One-against-all (OAA), which compares a given class with all the other classes.

According to a comparison study of Weston and Watkins (Weston and Watkins 1998), the accuracy of these methods is almost the same. Hsu and Lin (2002) gave a detailed comparison of different methods for the multi-class SVM on large experimental problems and concluded that One-Against-One method is a competitive approach and may be more suitable for practical use.

5 Classification Results and Discussions

In the present work, we have several types of defects, so it is important to not only detect these defects (detection stage) but also to classify them (identification stage). For this, a SVM classifier is specifically used at each stage of diagnosis: the detection stage, where the training set consists only of examples in normal and fault conditions (2 classes). The identification stage, where the training set consists only of examples in fault conditions (5 classes).

Here, we have applied the one-against-one approach for the multi-class classification using 10-fold CV. Where N is the number of classes, $N(N-1)/2$ classifiers are constructed and each one trains data from two classes. In the classification, we use a voting strategy in which each binary classification is considered to be a voting, where votes could be casted for all data points, x, at the end a point is designated to be in a class with the maximum number of votes (Widodo and Yang 2007; Chapelle et al. 1999; Weston and Watkins 1998; Hsu and Lin 2002). The classification accuracy is the percentage of number of correctly predicted data with respect to the total number of testing data. The 10-fold CV is used to reduce the bias related with random sampling of the training and test sets. The cross-validation accuracy (CVA) is the average of the k individual accuracy measures:

$$CVA = \frac{1}{k}\sum_{j=1}^{k} A_j \tag{9}$$

where k (10 in this case) is the number of folds used, and A_j is the accuracy measure of each fold, $j = 1,\ldots, k$.

The CVA of SVM classification in detection and identification stages, in normal, non-stationary and combined conditions, with and without selection, are shown respectively in Tables 4 and 5.

Table 4. Performance of SVM classification without selection

Conditions	Number of patterns		FV (17 indicators)	
				IAS1
Non-stationary	Load	300	Detection	100
		250	Identification	98.5
	Speed	60	Detection	100
		50	Identification	100
Normal (stationary)		900	Detection	100
		750	Identification	100
Combined conditions		1260	Detection	99.4
		1050	Identification	99.7

Table 5. Performance of SVM classification with DEFS selection

Conditions				IAS1
Non-stationary	Load	Detection	Success (%)	100
			Number of selected indicators	5
		Identification	Success (%)	100
			Number of selected indicators	4
	Speed	Detection	Success (%)	100
			Number of selected indicators	2
		Identification	Success (%)	100
			Number of selected indicators	2
Normal (stationary)		Detection	Success (%)	100
			Number of selected indicators	6
		Identification	Success (%)	100
			Number of selected indicators	5
Combined conditions		Detection	Success (%)	99.5
			Number of selected indicators	11
		Identification	Success (%)	99.8
			Number of selected indicators	12

The results show the performance of the classification for detection and identification stages. It appears clearly that

- In the normal condition, we remark that the used FV give perfect performances of 100%. In the case with selection, this performance is reached with only 6 and 5 indicators respectively for detection and identification stages.
- In the nonstationary conditions:
  - Load variation, the diagnosis success reaches a value of 100% for all cases and achieve 98.5% for identification stage. The selection improves the performance and achieves a 100% perfect accuracy to detect gear faults with only four optimal features selected from DEFS method.
  - Speed variation, in both stages, the accuracy of classification reaches 100% in all cases and particularly with only two selected indicators.
- In the combined conditions, it is found that the designed SVM classifier can diagnose all gear faults accurately, reaching respectively a success of 99.4% and 99.7% for detection and identification stages. The selection improves slightly the results, achieving a success of 99.5% for detection and 99.8% for identification with reduced feature subsets composed of 11 and 12 indicators respectively.

6 Conclusion

This paper presents a methodology for diagnosis of multiple gear failures under stationary, nonstationary and combined operating conditions. The presented methodology is based on the use of Instantaneous Angular Speed (IAS), Differential Evolution Feature Selection DEFS and multiclass SVM. It can be applied in the monitoring domain where rotating machines operate under variable speeds and are subjected to multiple failure modes.

According to the performances achieved without selection, it can be concluded that the IAS signals deliver highly relevant indicators whether in the stationary, non-stationary or combined case. On the other hand, the DEFS selection offers also the possibility to eliminate unnecessary indicators, improves the success rate and reduces CPU time. Consequently, using angular domain features extracted from IAS is highly recommended to diagnosis the multiple gear faults in all operating conditions.

The proposed approach is suitable for the monitoring of rotating machines operate under nonstationary condition. In future, the approach can be generalized by considering multiple failure modes and variable conditions.

Acknowledgments. This work was achieved at the laboratories LaMCoS (INSA - Lyon, France) and LMPA (IOMP, Sétif -1- University, Algeria). The authors would like to thank the Algerian and French Ministries of Higher Education and Scientific Research for their financial and technical support in the framework of program PROFAS 2012.

References

Bordoloi, D.J., Tiwari, R.: Optimum multi-fault classification of gears with integration of evolutionary and SVM algorithms. Mech. Mach. Theory **73**, 49–60 (2014)

Boulenger, A., Pachaud, C.: Aide-mémoire. Surveillance des machines par analyse des vibrations. Dunod, Paris (2009)

Chapelle, O., Haffner, P., Vapnik, V.N.: Support vector machines for histogram-based classification. IEEE Trans. Neural Netw. **10**(5), 1055–1064 (1999)

Dorigo, M., Stützle, T.: Ant Colony Optimization. MIT Press, London (2004)

Dubuisson, B.: Diagnostic et reconnaissance des formes. Traité des nouvelles technologies. Série: Diagnostic et Maintenance. Hermès, Paris (1990)

Fedala, S., Felkaoui, A., Zegadi, R., Ziani, R.: Optimisation des paramètres du vecteur forme: application au diagnostic vibratoire automatisé des défauts d'une boite de vitesse d'un hélicoptère. J. Matériaux Tech. **2**(97), 149–155 (2009)

Fedala, S., Rémond, D., Zegadi, R., Felkaoui, A.: Contribution of angular measurements to intelligent gear faults diagnosis. J. Intell. Manuf. (2015). https://doi.org/10.1007/s10845-015-1162-1

Fedala, S., Rémond, D., Zegadi, R., Felkaoui, A.: Gear fault diagnosis based on angular measurements and support vector machines in normal and nonstationary conditions. In: Chaari, F., et al. (ed.) Advances in Condition Monitoring of Machinery in Non-Stationary Operations. Applied Condition Monitoring 4, pp. 291–308 (2016). https://doi.org/10.1007/978-3-319-20463-5_22

Fyfe, K.R., Munck, E.D.S.: Analysis of computed order tracking. Mech. Syst. Signal Process. **11** (2), 187–205 (1997)

Hastie, T., Tibshirani, R., Friedman, J.: The Elements of Statistical Learning: Data Mining, Inference and Prediction, 2nd edn. Springer, New York (2009)

Haupt, R.L., Haupt, S.E.: Practical Genetic Algorithms. Wiley, New York (2004)

Hsu, C.W., Lin, C.J.: A comparison of methods for multi-class support vector machines. IEEE Trans. Neural Netw. **13**(2), 415–425 (2002)

Kennedy, J., Eberhart, R.C., Shi, Y.: Swarm Intelligence. Morgan Kaufman Publishers, London (2001)

Khushaba, R.N., Al-Ani, A., Al-Jumaily, A.: Feature subset selection using differential evolution and a statistical repair mechanism. Expert Syst. Appl. **38**(9), 11515–11526 (2011)

Kong, X.: Gear train monitoring by transmission error method. Ph.D. Thesis, University of Wisconsin-Madison (1987)

Kudo, M., Sklansky, J.: Comparison of algorithms that select features for pattern classifiers. Pattern Recogn. **1**(33), 25–41 (2000)

Lamraoui, M., Thomas, M., El Badaoui, M., Girardin, F.: Indicators for monitoring chatter in milling based on instantaneous angular speeds. Mech. Syst. Signal Process. **44**, 72–85 (2014)

Leclère, Q., Hamzaoui, N.: Using the moving synchronous average to analyze fuzzy cyclostationary signals. Mech. Syst. Sig. Process. **44**, 149–159 (2014)

Li, Y., Gu, F., Harris, G., et al.: The measurement of instantaneous angular speed. Mech. Syst. Sig. Process. **19**, 786–805 (2005)

Mohanty, A.R.: Machinery Condition Monitoring, Principles and Practices. CRC Press, Taylor & Francis Group, Boca Raton (2015)

Price, K., Storn, R.M., Lampinen, J.A.: Differential Evolution: A Practical Approach to Global Optimization (Natural Computing Series). Springer, Berlin (2005)

Randall, B.: Vibration-Based Condition Monitoring: Industrial. Wiley, New York (2011)

Rémond, D.: Practical performances of high-speed measurement of gear transmission error or torsional vibrations with optical encoders. Meas. Sci. Technol. **9**, 347–353 (1998)

Renaudin, L., Bonnardot, F., Musy, O., et al.: Natural roller bearing fault detection by angular measurement of true instantaneous angular speed. Mech. Syst. Sig. Process. **24**, 998–2011 (2010)

Samanta, B.: Gear fault detection using artificial neural networks and support vector machines with genetic algorithms. Mech. Syst. Sig. Process. **18**, 625–644 (2004)

Samanta, B., Al-Balushi, K.R., Al-Araimi, S.A.: Bearing fault detection using artificial neural networks and genetic algorithm. J. Appl. Sig. Process. **3**, 366–377 (2004)

Vachtsevanos, G., Lewis, F.L., Roemer, M., et al.: Intelligent Fault Diagnosis and Prognosis for Engineering Systems. Wiley, New Jersey (2006)

Vapnik, V.N.: The Statistical Learning Theory. Wiley, New York (1998)

Weston, J., Watkins, C.: Multi-class support vector machines. Technical Report CSD-TR-98-04. Royal Holloway University of London (1998)

Widodo, A., Yang, B.: Support vector machine in machine condition monitoring and fault diagnosis. Mech. Syst. Sig. Process. **6**, 2560–2574 (2007)

Effect of Input Data on the Neural Networks Performance Applied in Bearing Fault Diagnosis

Hocine Fenineche[1,2](✉), Ahmed Felkaoui[1], and Ali Rezig[3]

[1] Optic and Precision Mechanics Institute, University of Setif1, Setif, Algeria
fenihocine@yahoo.fr, a_felkaoui@yahoo.fr
[2] Department of Mechanical Engineering, University of Jijel, Jijel, Algeria
[3] Department of Electrical Engineering, University of Jijel, Jijel, Algeria
Ali.rezig@gmail.com

Abstract. The aim of this paper is to study the effect of input parameters choice of the artificial neural network (ANN), in order to obtain the best performances of fault classification. The purpose of this network is to automate the electric motor bearing diagnosis based on vibration signal analysis. The choice of the components of ANN's inputs (training and testing) has a big challenge for prediction of the machines faults diagnosis. The vibration signals collected from the test rig (Bearing Data Center) are preprocessed, to extract the most appropriate monitoring indicators to analyze the health of the experimental device.

To improve the performance of the neural network, we use three different dataset: the first contains only time indicators, while the second contains the frequency indicators, and the third set is a combination of these two indicators. A comparison between the effects of each feature on the ANN performances, allowed us to choose the optimal structure of input data. The obtained results show that the combined dataset give the best performances compared to the two others dataset.

Keywords: Artificial neural networks · Diagnosis · Bearing faults
Vibration analysis

1 Introduction

Bearings are the most fragile components of rotating machines. Being located between the fixed part and the moving part of these machines, they ensure the transmission of forces and the rotation of the shaft. They must be continuously monitored and any defect should be tracked to avoid costly production downtime.

However, the vibration signals generated by faults in such systems have been widely studied (McFadden and Smith 1985), and there are many signal processing techniques that can be used to extract the defect information from a measured vibration signals (Randall and Antoni 2011; Rai and Upadhyay 2016).

A. Felkaoui et al. (Eds.): SIGPROMD'2017, ACM 12, pp. 34–43, 2019.
https://doi.org/10.1007/978-3-319-96181-1_3

The artificial neural networks by their capacities of training, classification, and decision, give a solution to the problems of diagnosis bearings by the automatic classification of the vibratory signals, which corresponds to the various states of normal and abnormal functioning of the machines (Alguindigue et al. 1993; Samanta and Al-Balushi 2003; Rajakarunakaran et al. 2008; Li et al. 2000). The artificial neural networks are intended to increase the precision (accuracy) and to reduce errors caused by subjective human judgments.

The accuracy of ANN model highly depends on the setting of network parameters, such as sufficient number of hidden layers, neurons within each layer, and learning rate, as well as activation function. Most of the research in this area suggest some methods to find optimal parameters setting of the neural network (McCormick and Nandi 1996; Giuliani et al. 1998; Jack and Nandi 2000; Al-Araimi et al. 2004; Abhinav and Ashraf 2007; Rao et al. 2012).

However, very little attention has been paid to the effect of the dataset structure used to training and testing the ANNs. Therefore, the main objective of this work is to study the effect of the components choice of the input vector on the performances of the artificial neural network, to be used as a diagnostic tool of bearing defects. Starting from the analysis of signals collected by vibration sensors of the bearing test rig, with the calculation of time indicators (kurtosis, Rms, or crest factor) and frequency indicators. Then, configure them to build the database which will be used for learning and testing the ANN, which will allow us to find the best network configuration (inputs, outputs and parameters), and subsequently to automate the decision on the possibility of the fault bearings.

2 Background

2.1 Rolling Element Bearings

The main components of rolling bearings are the inner ring; the outer ring, the rolling elements, and the cage (see Fig. 1). Typically, the inner ring of the bearing is mounted on a rotating shaft, and the outer ring is mounted in the stationary housing. The rolling elements may be balls or rollers. The balls in a ball bearing transfer the load over a very small surface (ideally, point contact) on the raceways (Randall and Antoni 2011).

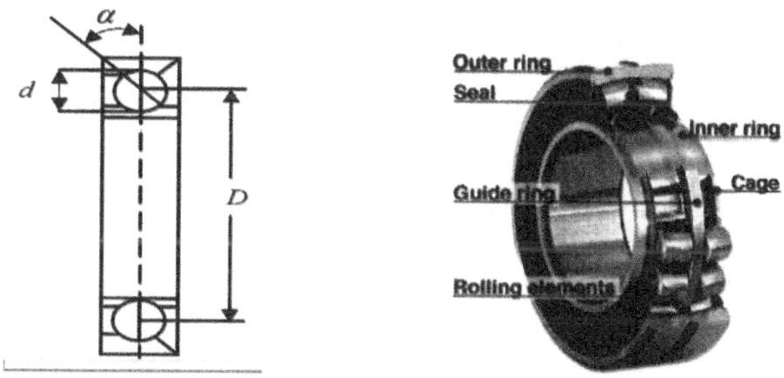

Fig. 1. Components of a rolling element bearing.

Local or wear defects causes periodic impulses in vibration signals. Amplitude and periodic of these impulses are determined by shaft rotational speed, fault location, and learning dimensions. The formula for the various defect frequencies is given by:

Ball pass frequency, outer race:

$$BPFO = \frac{nfr}{2}\left(1 - \frac{d}{D}\cos(\alpha)\right) \tag{1}$$

Ball pass frequency, inner race:

$$BPFI = \frac{nfr}{2}\left(1 + \frac{d}{D}\cos(\alpha)\right) \tag{2}$$

Fundamental train frequency (cage speed):

$$FTF = \frac{fr}{2}\left(1 - \frac{d}{D}\cos(\alpha)\right) \tag{3}$$

Ball (roller) spins frequency:

$$BSF = \frac{D}{2d}\left(1 - \left(\frac{d}{D}\cos(\alpha)\right)^2\right) \tag{4}$$

Where *fr* is the shaft speed, *n* is the number of rolling elements, and *f* is the angle of the load from the radial plane. Note that the ball spin frequency (BSF) is the frequency with which the fault strikes the same race (inner or outer).

2.2 Bearing Fault Diagnosis Technique

A wide variety of techniques based on various algorithms were developed for the detection and diagnosis of faults in rolling element bearings and have been introduced to inspect raw vibration signals. These algorithms can be classified into time domain, frequency domain, time-frequency domain, and higher order spectral analysis (Nataraj and Kappaganthu 2011).

2.3 Multi-Layer Perceptron (MLP)

The multi-layer perceptron (MLP) is the simplest and most known structure of the neural networks. This structure is shown in Fig. 2, is relatively simple with a layer of inputs, a layer of outputs and one or more hidden layers. Each neuron is not only connected to the neurons of the preceding layers, but also to all the neurons of the following layer (Bishop 1995).

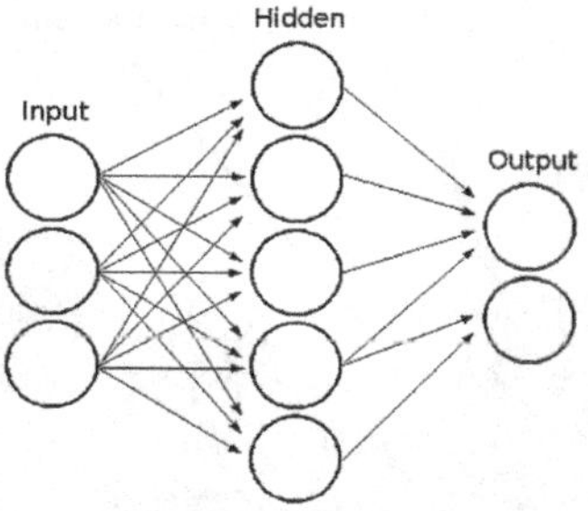

Fig. 2. Multi-layer perceptron general architecture.

The learning of the multilayer perceptron is supervised, and consists of adapting the weights of the neurons so that the network is capable of performing the requested task.

The conventional method for learning the multilayer perceptron is the back-propagation algorithm, which was developed in particular by Rumelhart and Parkenet le Cun in 1985. This algorithm relies on the minimization of the quadratic error between the computed outputs and those desired.

3 Materials and Methods

3.1 Data Acquisition

An experimental test rig built to predict the defects in rolling bearings is shown in Fig. 3.

This website provides access to ball bearing test data for normal and faulty bearings (Case Western Reserve University, bearing data Center 2006). Experiments were conducted using a 2 horsepower (hp) Reliance Electric motor, and acceleration data were measured at locations near to remote from the motor bearings. These web pages are unique in that the actual test conditions of the motor as well as the bearing fault status have been carefully documented for each experiment.

Motor bearings were seeded with faults using electro-discharge machining (EDM). Faults diameter ranging from 0.17 mm to 0.71 mm in diameter were introduced separately at the inner raceway, rolling element (i.e. ball) and outer raceway. Faults bearings were reinstalled into the test motor and vibration data were recorded for motor loads of 0 to 3 hp (motor speeds of 1797 to 1720 RPM). Vibration data were collected using accelerometers, which were attached to the housing with magnetic bases. Accelerometers were placed at the 12 o'clock position at both the drive end and fan end of the motor housing.

The time domain presentation of signal is shown in Fig. 4.

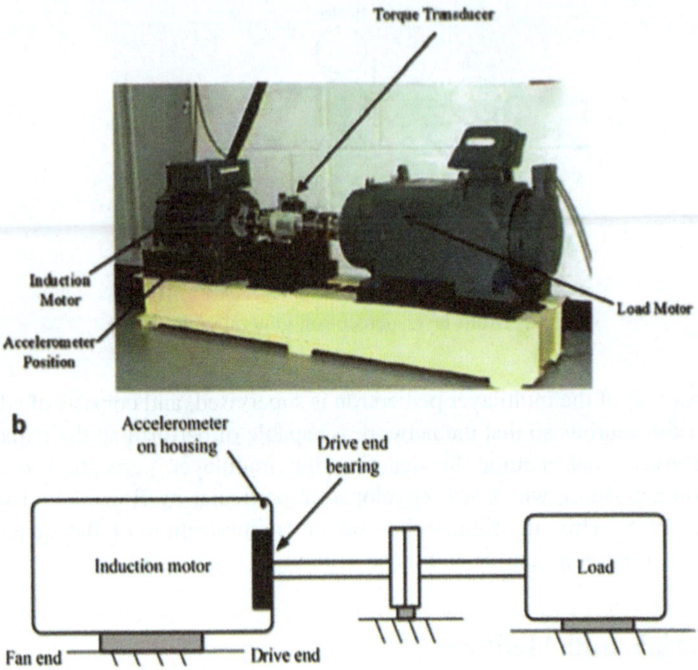

Fig. 3. (a) The bearing test rig; (b) the schematic description of the test rig. (Huang et al. 2010)

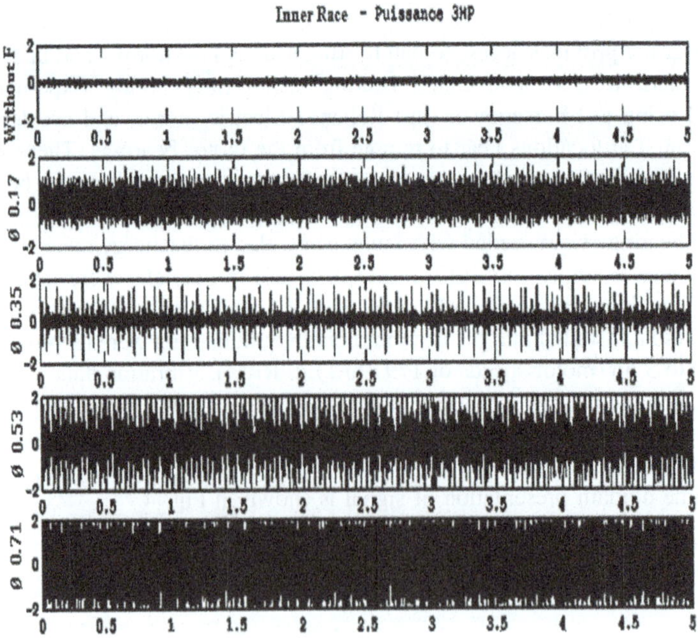

Fig. 4. The time domain signal

3.2 Preprocessing of Vibration Signals

A signal conditioning is required to remove all kinds of useless information, and to facilitate the task of extracting indicators for monitoring the most relevant formants database. We chose to calculate the following indicators: the root mean square value (RMS), crest factor, peak to peak value and kurtosis, and the energy from the spectrum envelope.

After a preliminary analysis (Fedala 2005), we choose to calculate these indicators as follows:

3.2.1 Time Domain Indicators

The time domain indicators (the root mean square value (RMS), crest factor, peak to peak value and kurtosis) are calculated in 5 frequency bands with a total width of 6000 Hz. Each of these 4 bands has a width of 1500 Hz, in addition to a total band that contains the four composed bands. The bands are then calculated within: [1–1500 Hz], [1500–3000 Hz], [3000–4500 Hz], [4500–6000 Hz], in addition to the total band of [1–6000 Hz]. The signal from each slice has been focused and filtered by a bandpass filter.

3.2.2 Frequencies Domain Indicators

As the same methodology used in the calculation of time domain indicators, the Frequencies domain indicator (the energy from the spectrum envelope) is calculated in five frequencies bands of a total width of 6000 Hz, in addition to the six large one that contain other bands with a total width of 6000 Hz. These bands are calculated as follows: [1–1000 Hz], [1000–2000 Hz], [2000–3000 Hz], [3000–4000 Hz], and [4000–5000 Hz], in addition to the total band of [1–6000 Hz].

3.3 Constitution of the Patterns Vector (Networks Input)

The patterns vector is consisted of three different dataset: the first contains only time indicators, while the second contains the frequency indicators, and the third set is a combination of these two indicators. As the main scope of this paper is limited to study the effect of the components choice of the input vector on the performances of the artificial neural network, to be used as a diagnostic tool of bearing defects. The detailed methodology of combining time domain indicator and frequency domain indicator can be found in the literature (Unal et al. 2014) 187–196 (Samanta and Al-Balushi 2003; Jack and Nandi 2002).

The data that must be classified and treated, are stored in an array of type observations/variables.

3.4 Choice of the Classes (Networks Output)

The network outputs vector contains various classes corresponding to each operating conditions from the experimental test rig. We chose five classes, each one of them corresponds to a diameter of the defect. Table 1 represents the labelling of the various studied classes.

Table 1. Labelling of the classes

Class	Fault diameter	Label
1	Without fault	10000
2	0.17 mm	01000
3	0.35 mm	00100
4	0.53 mm	00010
5	0.71 mm	00001

3.5 Data Standardization

To improve the performances of the MLP, it is preferable to normalize the data of the patterns vector. We divided the obtained database into two parts: a training set (70% of database) which train the network, while the remaining database (30%) were used for testing, on which, they have been presented to measure network's performances.

3.6 The Network Configuration

We used a multi-layer perceptron with the following configuration (Fenineche 2008):

- Only one hidden layer.
- 5 neurons in the hidden layer.
- 5 neurons in output layer which corresponding to the various classes.
- Performance Function: MSE (Mean Square Error).

4 Results and Discussion

Table 2 summarizes the values of the MSE error using the various indicators and parameters described above.

In each case, the network is trained until it reaches the values of the stop criteria. The results are obtained after several executions.

Table 2. Performance of the MLP classification

Indicator	MSE
Time	0.0324
Frequency	0.0320
Combined	0.0235

The Figs. 5, 6 and 7 show the performances of ANN for different input data. We have obtained a performance of 0.032 (for MSE) using the time indicators and a performance of 0.036 with the frequency indicators, while the combination of the two sets gives a better performance of 0.0235.

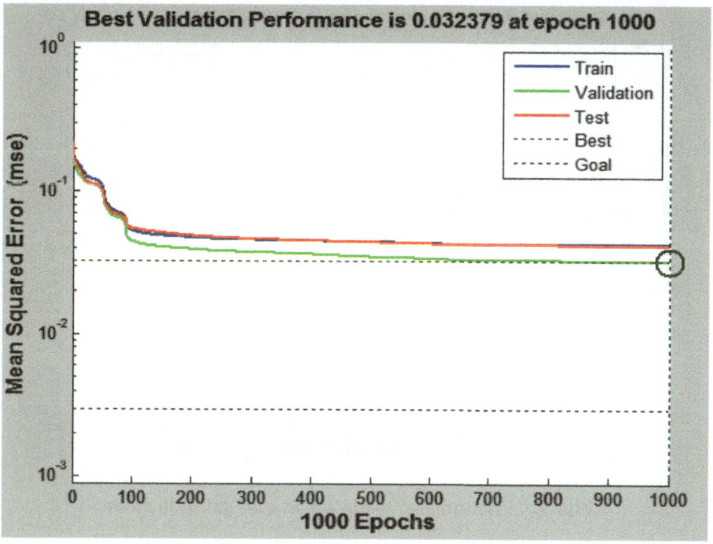

Fig. 5. Performance using the time indicators

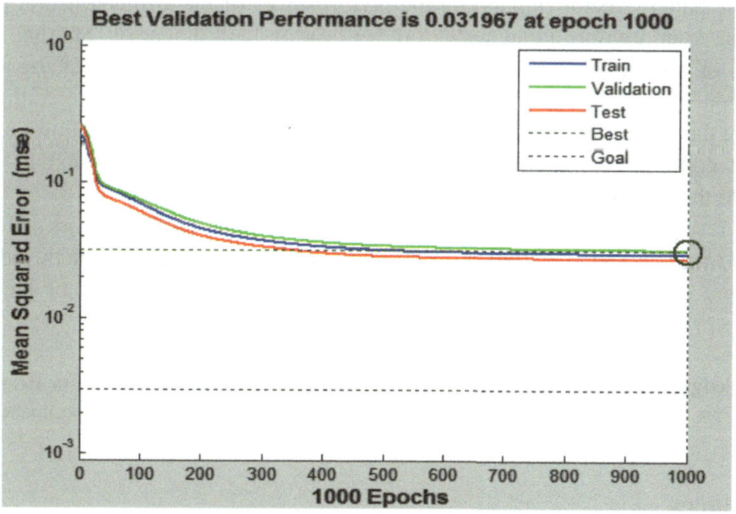

Fig. 6. Performance using the frequency indicators

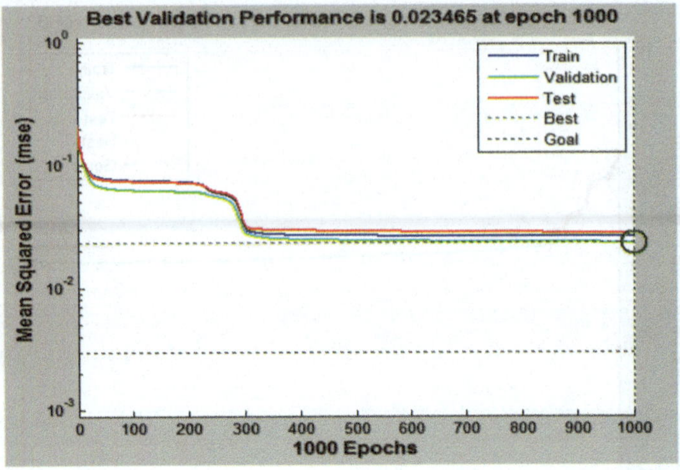

Fig. 7. Performance using the combined indicators

5 Conclusion

The objective of this work is to study the effect of the choice of the elements constituting the pattern vector (inputs) on the performances of the artificial neural network, which has been used as a diagnostic tool for bearing fault diagnosis. Starting from the analysis of the signals collected by vibration sensors of a rolling test rig, and the calculation of time indicators and frequency indicators. Then, they are configured to build the database that will be used to learn and test the ANN, which allows us to find the best configuration of the network (inputs, parameters and outputs) in order to automate the decision on the eventuality of a bearing defect.

The results show that the performance of the artificial neural network is better for the case with the combined indicators. This is because the combined data include all the indicators, which enable them to better presenting the health status of the studied system.

Acknowledgment. The authors would like to thank Kenneth A. Loparo, from Bearing Data Center, Case Western Reserve University, Cleveland, for providing us the experimental data.

References

McFadden, P.D., Smith, J.D.: The vibration produced by multiple point defects in a rolling element bearing. JSV **98**(2), 263–273 (1985)

Randall, R.B., Antoni, J.: Rolling element bearing diagnostics—A tutorial. Mech. Syst. Signal Process. **25**(2), 485–520 (2011)

Rai, A., Upadhyay, S.H.: A review on signal processing techniques utilized in the fault diagnosis of rolling element bearings. Tribol. Int. **96**, 289–306 (2016)

Alguindigue, I.E., Loskiewicz-Buczak, A., Uhrig, R.E.: Monitoring and diagnosis of rolling element bearings using artificial neural networks. IEEE Trans. Ind. Electron. **40**(2), 209–217 (1993)

Samanta, B., Al-Balushi, K.R.: Artificial neural network based fault diagnostics of rolling element bearings using time-domain features. Expert Syst. Appl. **17**, 317–328 (2003)

Rajakarunakaran, S., Venkumar, P., Devaraj, K., Rao, K.S.P.: Artificial neural network approach for fault detection in rotary system. ASC **8**(1), 740–748 (2008)

Li, B., Chow, M.Y., Tipsuwan, Y., Hung, J.C.: Neural network based rolling bearing fault diagnosis'. IEEE Trans. Ind. Electron. **47**(51), 1060–1067 (2000)

McCormick, A.C., Nandi, A. K.: Rotating machine condition classification using artificial neural networks. In: Proceedings of COMADEM 1996, University of Sheffield, 16–18 July 1996

Giuliani, G., Rubini, R., Maggiore, A.: Ball bearing diagnostics using neural networks. In: Proceedings of the Third International Conference Acoustical and Vibratory Surveillance Methods and Diagnostic Techniques. Senlis, France, pp. 767–776 (1998)

Jack, L.B., Nandi, A.K.: Feature selection for ANNs using genetic algorithms in detection of bearing faults. IEE Proc. Vision Image Signal Process. **147**(3), 205–212 (2000)

Al-Araimi, S.A., Al-Balushi, K.R., Samanta, B.: Bearing fault detection using artificial neural networks and genetic algorithm. EURASIP J. Adv. Signal Process. **2004**(3), 366–377 (2004)

Abhinav, S., Ashraf, S.: Evolving an artificial neural network classifier for condition monitoring of rotating mechanical systems. Appl. Soft Comput. **7**, 441–454 (2007)

Rao, B.K.N., et al.: Failure diagnosis and prognosis of rolling - element bearings using artificial neural networks: a critical overview. In: 25th International Congress on Condition Monitoring and Diagnostic Engineering. Journal of Physics Conference Series 364 (2012)

Nataraj, C., Kappaganthu, K.: Vibration-based diagnostics of rolling element bearings: state of the art and challenges. In: 13th World Congress in Mechanism and Machine Science, Guanajuato, Mexico, 19–25 June 2011

Bishop, C.M.: Neural Networks for Pattern Recognition, p. 498. Oxford University Press, Oxford (1995)

Huang, Y., Liu, C., Zha, X.F., Li, Y.: A lean model for performance assessment of machinery using second generation wavelet packet transform and fisher criterion. Expert Syst. Appl. **37**, 3815–3822 (2010)

Case Western Reserve University, bearing data Center (2006). http://www.eecs.cwru.edu/laboratory/bearing/download.htm. Case Western Reserve University, bearing data center

Fedala, S.: Le diagnostic vibratoire automatisé: comparaison des méthodes d'extraction et de sélection du vecteur forme, Magister thesis, University of Setif (2005)

Unal, M., Onat, M., Demetgul, M., Kucuk, H.: Fault diagnosis of rolling bearings using a genetic algorithm optimized neural network. Measurement **58**, 187–196 (2014)

Jack, L.B., Nandi, A.K.: Fault detection using support vector machines and artificial neural networks augmented by genetic algorithms. Mech. Syst. Signal Process. **16**(2–3), 373–390 (2002)

Fenineche, H.: Application des réseaux de neurones artificiels au diagnostic des défauts des machines tournantes. Magister Thesis, University of Setif (2008)

Bearing Diagnostics Using Time-Frequency Filtering and EEMD

Hafida Mahgoun(✉) and Ridha Ziani

LMPA Laboratory, Institute of Optics and Fine Mechanics, University of Ferhat Abbas, Setif 1, Algeria
mahafida006@yahoo.fr

Abstract. The ensemble empirical mode decomposition (EEMD) was largely used in the diagnosis of the rotating machines but the EEMD shows a limitation with the detection of the impulses that are influenced by the presence of noise, the mode mixing, and the end effect. To detect the shocks due to the defect at an early stage, we propose to use the Time-frequency filtering (TFF) which was recently proposed by Flandrin. This method allows us to denoise the signal and gives promising results in the detection of the defects on machine elements.

In this work first, we show by simulated bearing signal the advantage of TFF compared to the EEMD in the detection of impulses. Then, we analyze real vibration bearing signals by using the two different time-frequency methods, ensemble empirical mode decomposition (EEMD) and Time-frequency filtering (TFF), and then we compare the results given by using the two methods separately and the results by a new method when we combine the two methods. The filtered modes are analyzed by calculation of the spectrum, which gives more information about the defect and allows us to read it frequencies and detect it at an early stage.

Keywords: Time-frequency filtering (TFF) · Bearing · Fault detection
Empirical mode decomposition (EEMD) · Short fourier transformation
Time-frequency · Denoise

1 Introduction

Rolling bearings are widely used in rotating machines. Studies show that failure and breakdowns on rotating machines are generally related to bearings (Li and Ma 1997). Therefore, the fault must be detected as early as possible to avoid sudden breakdowns which lead to significant economic losses and human casualties.

Generally, a failure in a bearing component may begin as a manufacturing related defect or be induced by operating stress via overload or cyclic loading. Defects on bearing are classified to distributed defects and localized defects. Distributed defects include surface roughness, waviness, misaligned races and off-size rolling elements. Localized defects include cracks, pits and spalls on the rolling surfaces (Li and Ma 1997).

When a fault is occurring in a roller bearing, it often generates periodic impacts every time the roller element hits a defect in the raceway or every time a defect in the roller element hits the raceway (McFadden and Smith 1984), The size and the

A. Felkaoui et al. (Eds.): SIGPROMD'2017, ACM 12, pp. 44–55, 2019.
https://doi.org/10.1007/978-3-319-96181-1_4

repetition period of the impact are determined by the shaft rotation speed, the type of fault and the geometry of the bearing (McFadden and Smith 1985). These parameters influence directly the vibration signal which is usually, non-stationary and non-linear. The non-linearity is due to some factors such as loads, clearance, friction, stiffness and the effect of lubrication. Since the roller bearing vibration signal is non-linear and non-stationary, therefore, it is difficult to identify the bearing fault using conventional methods.

Many methods based on vibration signal analysis have been developed to analyze the roller bearing signals; the envelope analysis technique is widely used as a powerful tool in the detection and diagnosis of bearing faults (McFadden and Smith 1984). This technique consists of bandpass filter followed by a demodulation and a Fast Fourier Transformation. However, the major shortcoming of envelope analysis is that the selection of the centre frequency and the bandwidth of the filter, these parameters are based on the historical data and are determined with experience which will make a great subjective influence on the diagnosis results.

In order to overcome the limitation of the envelop analysis we use the TFF method proposed by Flandrin (2015), Meignen et al. (2016) to filter the signal to analyze the vibration bearing signals for different defects size, and then we compare the results given by using the two methods separately and the results when we combine the two methods.

The structure of the paper is as follows: Sect. 2 introduces the basics of EEMD. In Sect. 3 the time-frequency filtering algorithm is summarized. In Sect. 4, we present simulated signal which is analyzed by using the two methods EEMD and TFF. In Sect. 5, the TFF and EEMD methods are applied to bearing faults diagnosis, and a conclusion of this paper is given in Sect. 6.

2 Time-Frequency Filtering (TFF)

The Time-Frequency Filtering technique is based on the fact that the Short-Time Fourier Transform (STFT) of a signal is entirely characterized by its zeros (Flandrin 2015; Meignen et al. 2016). Flandrin (2015) propose to use an algorithm based essentially on Delenaury triangulation and spectrogram $S_x^h(t, \omega)$ given by:

$$S_x^h(t, \omega) = \left|X_x^h(t, \omega)\right|^2, \tag{1}$$

where $X_x^h(t, w)$ is the STFT of $x(t)$

$$X_x^h(t, \omega) = \int_{-\infty}^{+\infty} x(s)\overline{h(s-t)}exp(-i\omega(s-\frac{t}{2}))ds. \tag{2}$$

The spectrogram is characterized by it zeros which are determined by using Weitrass-Hadamard factorization where time and frequency are considered as coordinates of a complex-valued variable by introducing $z = \omega + it$ and $X_x^h(t,\omega)$ can be written as:

$$X_x^h(t,\omega) = \exp\left(-\frac{1}{4}|z|^2\right)\varkappa_x(z), \tag{3}$$

for a Gaussian window and if we take

$$X_x(z) = \int_{-\infty}^{+\infty} A(z,s)x(s)ds, \tag{4}$$

and $A(z,s)$ is defined as:

$$A(z,s) = \pi^{-\frac{1}{4}}exp\left(-\frac{1}{2}s^2 - isz + \frac{1}{4}z^2\right) \tag{5}$$

The Time-Frequency Filtering algorithm is summarized as follows:

(1) Perform Delaunay triangulation over STFT zeros z_m
(2) Identify outlier edges such that

$$|e_{mn}| = d(z_m, z_n) > \rho = 2. \tag{6}$$

(3) Keep triangles with at least one outlier edge
(4) Group adjacent such triangles in connected, disjoint domains D_j
(5) Multiply STFT with labelled 1/0 masks $1_{D_j}(t,\omega)$
(6) Reconstruct the disentangled components, domain by domain.

3 EMD and EEMD Algorithms

The EMD (Huang et al. 1998) consists to decompose iteratively a complex signal into a finite number of intrinsic mode functions (IMFs) which verify the two following conditions:

The number of extrema and the number of zeros of an IMF must be equal or differ at most by one.

An IMF must be symmetric with respect to local zero mean.

For a given signal $x(t)$ the EMD algorithm used in this study is given in literatures (Huang et al. 1998).

To alleviate the mode mixing effect of EMD, the EEMD was used. The EEMD decomposition algorithm of the original signal $x(t)$ used in this work is summarized in the following steps (Wu and Huang 2009):

Add a white noise $n(t)$ with given amplitude β_k to the original signal $x(t)$ to generate a new signal:

$$x_k(t) = x(t) + \beta_k n(t). \tag{7}$$

Use the EMD to decompose the generated signals $x_k(t)$ into N IMFs $IMF_{nk}(t)$, $n = 1, \ldots, N$, where the nth IMF of the kth trial is $IMF_{nk}(t)$.

Repeat steps (1) and (2) K times with different white noise series each time to obtain an ensemble of IMFs: $IMF_{nk}(t)$, $k = 1, \ldots, K$.

Determine the ensemble mean of the K trials for each IMF as the final result:

$$IMF_n(t) = \lim_{k \to \infty} \frac{1}{K} \sum_{k=1}^{K} IMF_{nk}(t), \quad n = 1, \ldots, N. \tag{8}$$

4 Simulation

In order to confirm the validity of the TFF and to compare the two methods, EEMD and TFF, we suggest analyzing a bearing simulated signal. This comparison will highlight the advantages of applying TFF in fault diagnosis.

In our study we will use a mathematical model suggested by Yuh-Tay Sheen (2004), to simulate the vibratory signal of a bearing with a defect in the inner ring (Fig. 1). The results of this model give a signal very similar to the real vibratory signal. The simulated signal composed of simulated pulse train exponentially decaying impulses. Each pulse is modulated by three signal harmonic frequencies with an exponential decay (Fig. 1).

The vibration signal is given by (Sheen 2004):

$$x(k) = e^{-\alpha kt}(sin2\pi f_1 kT + sin2\pi f_2 kT), \tag{9}$$

where $t = mod(kT, 1/f_o)$

$\alpha = 800$, $f_0 = 100\,\text{Hz}$, $f_1 = 3000\,\text{Hz}$,

$f_2 = 8000\,\text{Hz}$ and the sampling frequency is $fe = 25\,\text{kHz}$ so $T = 1/25000\,\text{s}$.

The same signal with additive noise is shown in (Fig. 2) the impulses are buried in noise. The spectrum given in Fig. 3 does not give clear information about the frequency defect.

By using EEMD, the noisy signal is decomposed using the ensemble number of 100. The first IMF is shown in Fig. 4. From this figure, it can be noticed that this first IMF corresponds to the transient component, by using the EEMD method, it is shown that the pulse repetition frequency is extracted efficiently and noise is retained in the first and the second IMF. Figure 4 shows also the first IMF filtered by using the TFF method, we can see that the noise is completely removed but also some information was lost. This information can be retrieved by using the supplementary IMFs.

By using the TFF, the noisy signal $x_s n(t)$ is also filtered (Fig. 5) and then decomposed. If we compare the results, we can observe that the first IMF (transient component) given by EEMD is noisy, because the noise used in the EEMD algorithm occur generally in the first IMF, we can observe that this noise does not exist in the filtered signal (Fig. 5) but also some information was lost by using only TFF.

To increase the information, in this study we propose to use the other IMFs in the reconstructed signal (Fig. 6) and to obtain the filtered signal by using the two methods (EEMD and TFF). From Fig. 6 we can see that the noisy signal is filtered and the information is conserved.

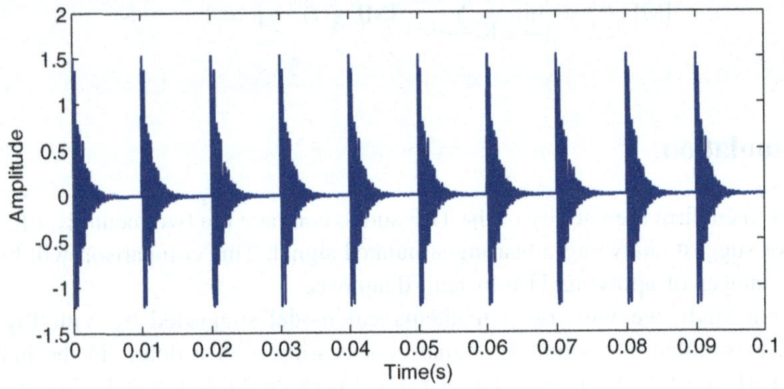

Fig. 1. Bearing simulated signal.

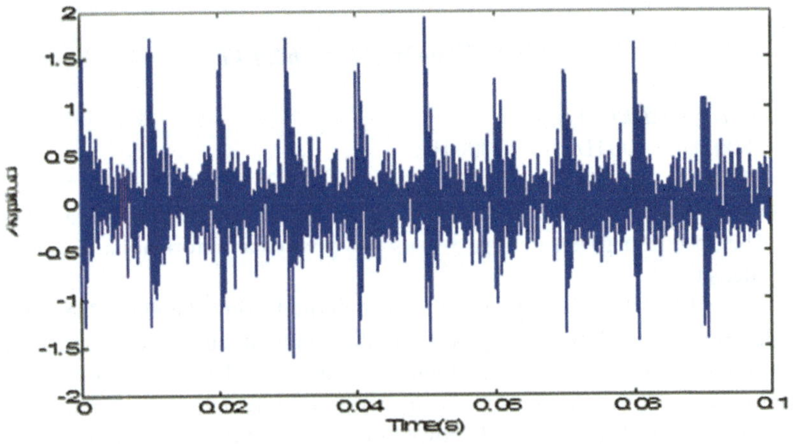

Fig. 2. Bearing simulated signal with additive noise.

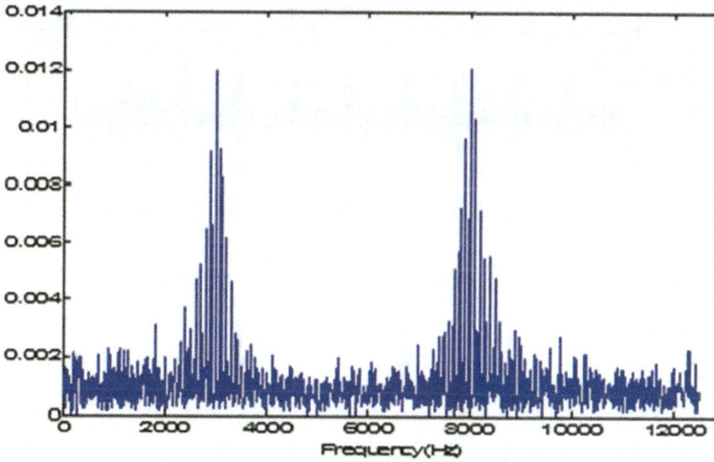

Fig. 3. The spectrum of the noisy signal.

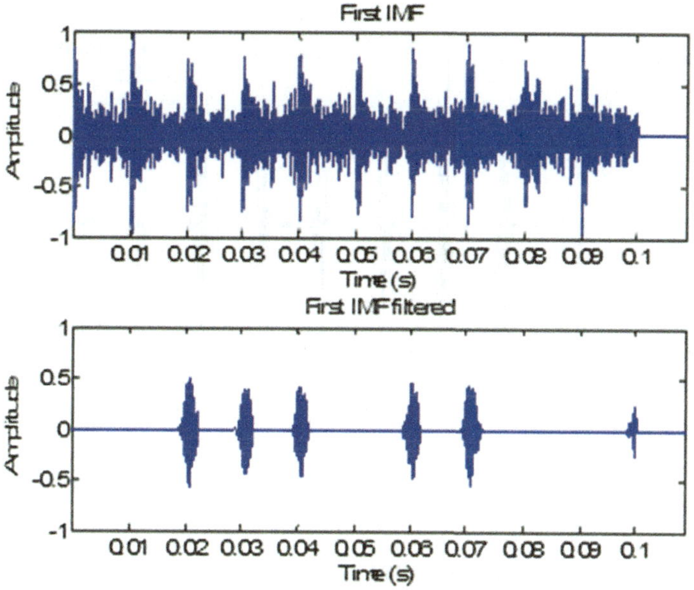

Fig. 4. First IMF filtered.

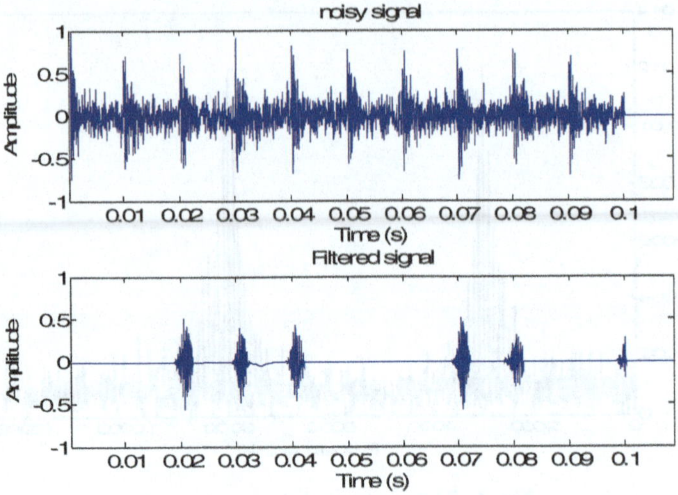

Fig. 5. The filtered signal using TFF.

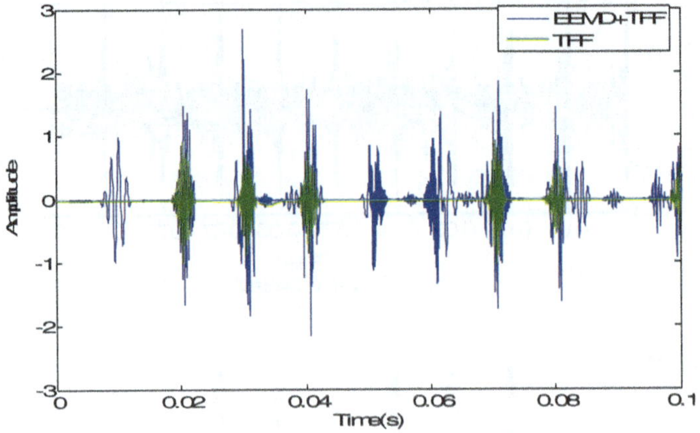

Fig. 6. The filtered signal using only TFF and (EEMD + TFF).

5 Application to Experimental Data

Experimental data (www.eecs) was collected from the drive-end ball bearing of an induction motor (Reliance Electric 2HP IQPreAlert)-driven mechanical system is shown in Fig. 7. The motor was connected to a dynamometer and a torque sensor by a self-aligning coupling. The dynamometer is controlled so that desired torque load levels can be achieved (www.eecs). The accelerometer was mounted on the motor housing at the drive end of the motor. The data collection system consists of a high bandwidth amplifier particularly designed for vibration signals. The accelerometer was mounted on the motor housing at the drive end of the motor. The data collection system consists of a high bandwidth amplifier particularly designed for vibration signals and a data recorder with a sampling frequency of 12,000 Hz per channel. The data recorder is equipped with low-pass filters at the input stage for anti-aliasing. The bearings were running at approximately 1797. The bearings used were SKF 6205 bearings. Some parameters are listed in Table 1: Size: (inches).

Table 1. Bearing parameters.

Parameters	Dimensions (inch)
Inside diameter	0.9843
Outside diameter	2.0472
Thickness	0.5906
Ball diameter	0.3126

Fig. 7. Test bench (the Case Western Reserve University- Bearing Data Center) (Kenneth 2003)

Table 2. Bearings fault frequencies.

Parameters	Frequency (Hz)
Inner race	162.3
Outer race	107
FFTE	12
Rolling element	141.36

The bearing fault frequencies (Table 2) can be computed as a function of the bearing geometry and of the operating speed (Kenneth 2003).

$$f_c = \frac{1}{2}\left(1 - \frac{d}{D_m} cos\alpha\right) f_r. \tag{10}$$

$$f_i = \frac{z}{2}\left(1 + \frac{d}{D_m} cos\alpha\right) f_r. \tag{11}$$

$$f_b = \frac{D_m}{d}\left(1 - \left(\frac{d}{D_m}\right)^2 cos\alpha\right) f_r. \tag{12}$$

$$f_o = \frac{z}{2}\left(1 - \frac{d}{D_m} cos\alpha\right) f_r, \tag{13}$$

where f_c is the cage fault frequency, f_i is the inner raceway fault frequency, f_o *is* the outer raceway fault frequency, f_b is the ball fault frequency, d is the ball diameter, Dm is the pitch diameter, Z is the number of rolling elements, fr is the rotation frequency of the shaft and is the ball contact angle.

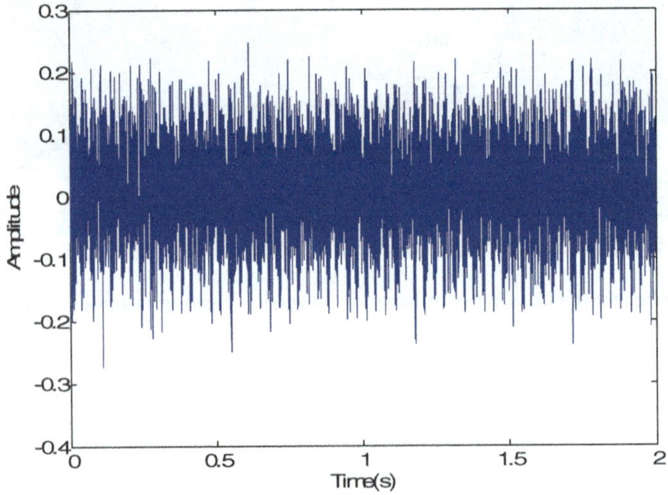

Fig. 8. Vibration signal from a normal roller bearing.

Figure 8 displays the vibration signal from a normal roller bearing which is analyzed with the EEMD and TFF. The first filtered IMF is displayed in Fig. 9. The spectrum of the filtered IMF is displayed in Fig. 10. We can see only and clearly the rotation frequency and its harmonics.

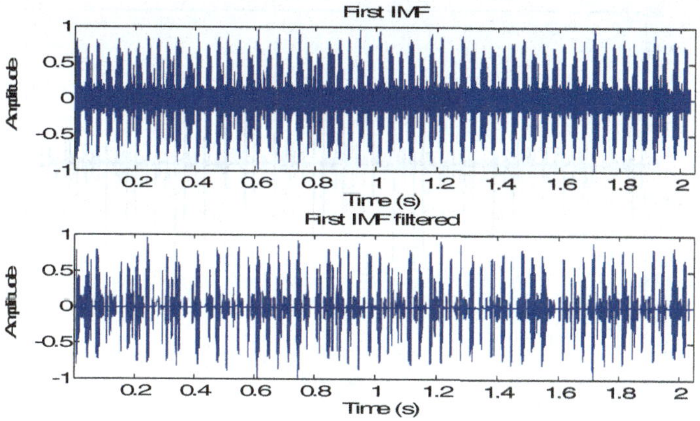

Fig. 9. First IMF filtered.

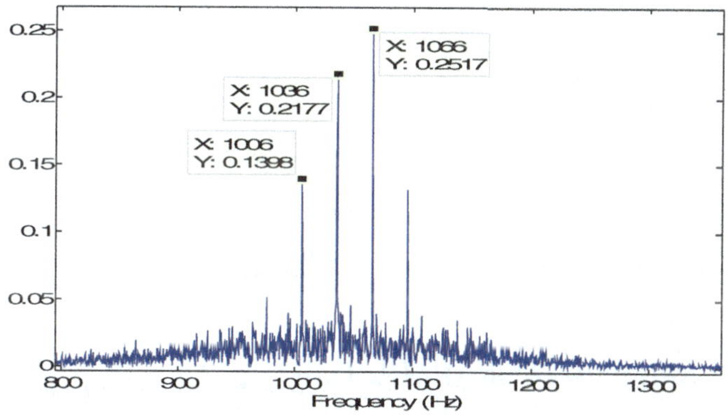

Fig. 10. IMF1 filtered spectrum.

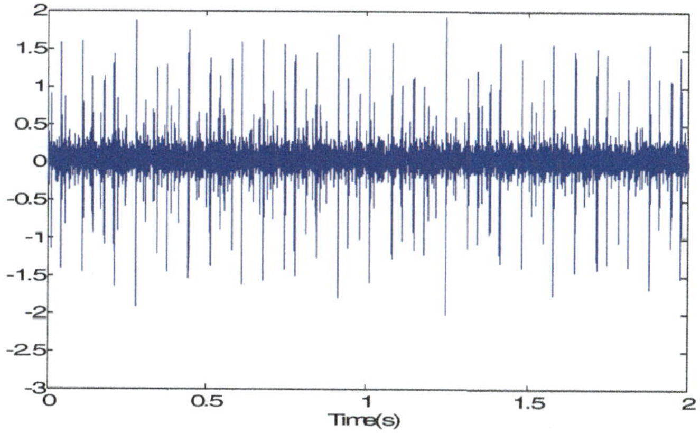

Fig. 11. Vibration signal from a faulty roller bearing

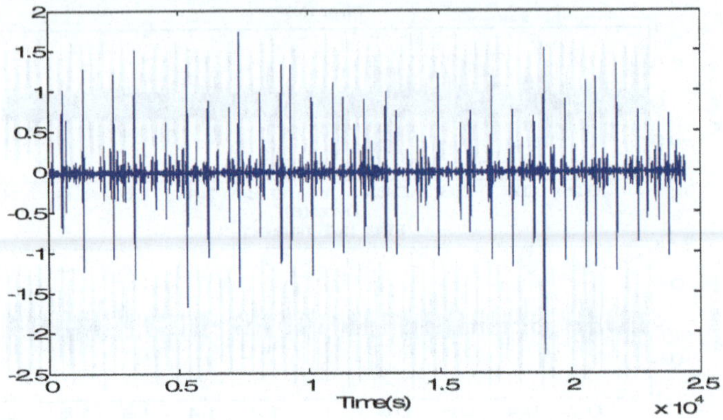

Fig. 12. The filtered signal from a faulty roller bearing.

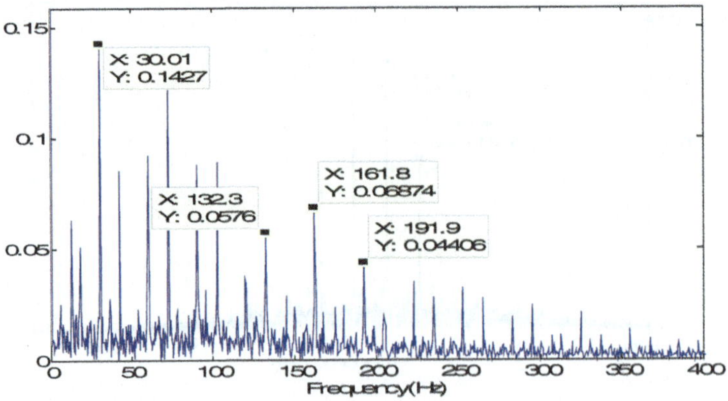

Fig. 13. The spectrum of the filtered signal from a faulty roller bearing.

The vibration signal of inner race defect is displayed in Fig. 11. We can see that the amplitude is very large and there are many impulses due to defect and at the same time we can see a considerable noise. The filtered signal using EEMD and TFF is shown in Fig. 12. It is clear that there are periodic impacts in the vibration signal. There are significant fluctuations in the peak amplitude of the signal, and there are also considerable variations of frequency content.

Figure 12 shows the results given by the EEMD and TFF. We can see from the spectrum Fig. 13 clearly the ball pass frequency of the inner race BPFI = 162 Hz and its harmonics, so it confirms the existence of the inner race fault.

6 Conclusion

In this study, we have combined the two methods EEMD and TFF to denoise and to analyze bearing vibration signals. The technique achieves good frequency separation and does not require the use of the envelope analyses.

The method has shown successful separation of the different modes that correspond to the presence of a defect. We have also used the spectrum to detect the frequency defect, and we have observed that the technique separates the different parts of the signal, and gives a solution to the problem of envelope analysis, which is the selection of the centre frequency and the choice of the bandwidth of the filter that is based on the historical data and is generally determined with experience.

References

Li, C.J., Ma, J.: Wavelet decomposition of vibrations for detection of bearing localized defects, nondestructive testing and evaluation (NDT&E). International **30**(3), 143–149 (1997)

McFadden, P.D., Smith, J.D.: The vibration produced by multiple point defects in a rolling element bearing. J. Sound Vib. **98**(2), 263–273 (1985)

McFadden, P.D., Smith, J.D.: Vibration monitoring of rolling element bearings by the high-frequency resonance technique. A review. Tribol. Int. **17**, 3–10 (1984)

Flandrin, P.: Time-frequency filtering based on spectrogram zeros. IEEE Signal Process. Lett. **22** (11), 2137–2141 (2015)

Meignen, S., Oberlin, T., Depalle, P., Flandrin, P., McLaughlin, S.: Adaptive multimode signal reconstruction from time-frequency representations. Phil. Trans. R. Soc. A **374**, 20150205 (2016). https://doi.org/10.1098/rsta.2015.0205

Huang, N.E., Shen, Z., Long, S.R.: The empirical mode decomposition and the Hilbert spectrum for nonlinear and non-stationary time series analysis. Proc. Royal Soc. Lond. Ser. **454**, 903–995 (1998)

Wu, Z., Huang, N.E.: Ensemble empirical mode decomposition: a noise-assisted data analysis method. Adv. Adapt. Data Anal. **1**, 11–41 (2009)

Sheen, Y.-T.: A complex filter for vibration signal demodulation in bearing defect diagnosis. J. Sound Vib. **276**, 105–119 (2004)

Kenneth, L.A.: Case Western Reserve University Bearing Data Center (2003). www.eecs.cwru.edu/laboratory/bearing/apparatus.htm

The Time-Frequency Filtering (TFF) Method Used in Early Detection of Gear Faults in Variable Load and Dimensions Defect

Hafida Mahgoun[1(✉)], Fakher Chaari[2], Ahmed Felkaoui[1], and Mohamed Haddar[2]

[1] Laboratoire de Mécanique de Précision Appliquée, Université Ferhat Abbas Sétif 1, Sétif, Algeria
hafida_2016@univ-setif.dz

[2] Laboratoire de Mécanique, Modélisation et Production (LA2MP), Ecole Nationale d'ingénieurs de Sfax, Sfax, Tunisia

Abstract. In stationary condition, a local gear fault is presented by periodic impulses. However, under variable load, the vibration signal is non-stationary and the periodic impulses are masked by the noise and the part of the signal due to the load. The use directly of the time-frequency methods doesn't allow detecting these impulses. In this study, we propose to use two different time-frequency methods, ensemble empirical mode decomposition (EEMD) and time-frequency filtering (TFF) to analyze the vibration signal. First, the EEMD method is used to decompose the vibration signals in many modes. Then each mode is filtered and denoised by using the TFF method. In this paper, we propose to compare the results given by using the two methods separately and the results when we combine the two methods.

Keywords: Fault detection · Variable speed
Empirical mode decomposition (EEMD) · Time-Frequency filtering (TFF)
Short time fourier transform (STFT) · Gear · Rotating machines

1 Introduction

Gears are mechanisms widely used for power transmission in rotating machinery. The malfunctions and defects of gears are inevitable. The faulty gear is usually the major source of noise and vibration (McFadden 1986) and may result in the abnormal operation and failure of the system. The early detection of gear faults is very important to prevent the system from damage.

When a local gear fault occurs, the vibration signal is characterized by the presence of periodic impulses (Mahgoun et al. 2016). If the rotating speed of the shaft is invariable. However, under the variable rotating speed of the shaft, and the vibration signal is non-stationary (Wu et al. 2012). The local gear fault induces periodic impulses but these impulses are masked. The use of the conventional methods of diagnosis of a defect such as the Fourier analysis, short time Fourier transform does not allow us to obtain good results. To avoid this problem, we propose to use time-frequency methods which

A. Felkaoui et al. (Eds.): SIGPROMD'2017, ACM 12, pp. 56–67, 2019.
https://doi.org/10.1007/978-3-319-96181-1_5

decompose the signal into bands and make it possible to filter the signal; these methods are very useful in these situations.

Until now, many time-frequency methods were applied to detect the fault at an early stage, among these methods such as Wigner Ville decomposition (WVD) (Forrester 1989), short Fourier Transform (STFT) (Staszeweski 1997) and wavelet transform (WT) (Wang and McFadden 1997) seem to be the suitable tool to identify the signal frequency and to provide information about the time variation of the frequency (Cohen 1989). These methods are classified to linear time-frequency representation such as STFT and wavelet transform, and bilinear methods such as Wigner Ville distributions. The STFT is appropriate only for analyzing signals with slow variation (Mallat 1998) and it is inefficient in the case of non-stationary signals. The WT transform was widely applied because it's a multiresolution analysis (Mallat 1998), is very used to detect the transient features to extract impulses and for denoising. Nevertheless, the wavelet analysis is also a linear transform and it uses functions named wavelet as window function like the STFT. The window changes its width by using a dilatation parameter. Then, at the high frequency, we have high time resolution and a low frequency resolution. While at low frequencies, we have low time resolution and high-frequency resolution. Then, we can't have a good resolution for all time-scale map due to the Heisenberg uncertainty principle (Staszewski 1997). In addition, this method gives a time-scale representation which is difficult to interpret as a time-frequency representation; we must have a relation between the scale and the frequency to understand the obtained results and to identify the fault frequencies. Another limitation of the WT is how to select the mother wavelet used in the analyses of the signal since different wavelets have different time-frequency structures (Peng and Chu 2004), also, how to calculate the range scale used in the WT is another deficiency of the transform (Liu et al. 2006). Many researchers demonstrated that the use of the WT introduces border distortion and energy leakage (Peng and Chu 2004).

In mechanical application, Yang et al. (2011) confirm that this method is highly dependent on the rotational speed and pre-knowledge of the machine. To overcome the deficiencies of these methods empirical mode decomposition (EMD) was proposed by Huang et al. (1998) for nonlinear and non-stationary signals and was applied in fault diagnosis of rotating machinery (Liu et al. 2006; Mahgoun et al. 2010). It does not use a priori determined basis functions and can iteratively decompose a complex signal into a finite number of zero mean oscillations named intrinsic mode functions (IMFs). Each resulting elementary component (IMF) can represent the local characteristic of the signal. However, one of the problems of EMD is mode mixing as a result of intermittency (Huang et al. 2003; Rilling and Flandrin 2008). Mode mixing occurs when different frequencies that should appear separately in different IMFs are presented in one IMF. This problem gives a vague physical significance of the IMF. EMD is unable to separate different frequencies in separate IMFs. Also, the IMFs are not orthogonal each other, which produce end effects. To solve the problem of mode mixing the ensemble empirical mode decomposition EEMD method was proposed by Wu and Huang (2009) by adding several realizations of Gaussian white noise to the signal, and then using the EMD to decompose the noisy signal, multiple IMFs can be obtained and the added noise is canceled by averaging the IMFs. The ensemble empirical mode decomposition (EEMD) proposed by Huang et al. to analyze nonlinear and non-stationary signals. The method was largely applied in fault diagnosis of rotating machinery (Mahgoun et al. 2012; Wu and Chung 2009) because it does not use a priori determined basis functions and can iteratively decompose a complex signal into a finite number of intrinsic mode functions (IMFs). Each resulting elementary component IMF can represent the local characteristic of the signal. We have used the EEMD to analyze non stationary signals collected from test bench which work under non stationary conditions, where the speed of the shaft is variable due to a variation of the load and we have obtained good results in impulses detection (Mahgoun et al. 2016) but we cannot separate the part caused by the variation of load and the part of the signal (impulses) due to defect.

In this work, we use two different time-frequency methods, ensemble empirical mode decomposition (EEMD) and time-frequency filtering (TFF) (Flandrin 2015; Meignen et al. 2016) to analyze the vibration signals given by a dynamic modeling of a gear transmission in the case of non stationary load and speed with a variation in the defect size, and then we compare the results given by using the two methods separately and the results when we combine the two methods.

The structure of the paper is as follows: In Sect. 2 the time-frequency filtering algorithm is summarized, Sect. 3 introduces the basic EMD and EEMD. In Sect. 4, the methods are applied and results are compared. In Sect. 5, a conclusion of this paper is given.

2 Time-Frequency Filtering (TFF)

The Time frequency filtering technique is based on the fact that the Short-Time Fourier Transform (STFT) of a signal is entirely characterized by its zeros (Flandrin 2015; Meignen et al. 2016). Flandrin (2015) proposes to use an algorithm based essentially on Delenaury triangulation and spectrogram $S_x^h(t,\omega)$ given by:

$$S_x^h(t,\omega) = \left|X_x^h(t,\omega)\right|^2 \tag{1}$$

Where $X_x^h(t,w)$ is the STFT of $x(t)$:

$$X_x^h(t,\omega) = \int_{-\infty}^{+\infty} x(s)\overline{h(s-t)}\exp(-i\,\omega\left(s-\frac{t}{2}\right))ds \tag{2}$$

the spectrogram is characterized by it zeros which are determined by using Weitrrass-Hadamard factorization where Time and frequency are considered as coordinates of a complex-valued variable by introducing $z = \omega + it$ and $X_x^h(t,\omega)$ can be written as:

$$X_x^h(t,\omega) = \exp\left(-\frac{1}{4}|z|^2\right)\varkappa_x(z) \tag{3}$$

For a Gaussian window and with

$$\varkappa_x = \int_{-\infty}^{+\infty} A(z, s)x(s)ds \tag{4}$$

and

$$A(z, s) = \pi^{-\frac{1}{4}}\exp\left(-\frac{1}{2}s^2 - isz + \frac{1}{4}z^2\right) \tag{5}$$

The time-frequency filtering algorithm is summarized as follows:

(1) Perform Delaunay triangulation over STFT zeros z_m
(2) Identify outlier edges such that

$$|e_{mn}| = d(z_m, z_n) > \rho = 2 \tag{6}$$

(3) Keep triangles with at least one outlier edge
(4) Group adjacent such triangles in connected, disjoint domains D_j
(5) Multiply STFT with labelled 1/0 masks $1_{Dj}(t, \omega)$
(6) Reconstruct the disentangled components, domain by domain.

3 EMD and EEMD Algorithms

3.1 EMD Algorithm

The EMD consists to decompose iteratively a complex signal into a finite number of intrinsic mode functions (IMFs) which verify the two following conditions:

(a) The number of extrema and the number of zeros of an IMF must be equal or differ at most by one.
(b) An IMF must be symmetric with respect to local zero mean.

For a given signal $x(t)$ the EMD algorithm used in this study is given in literatures (Huang 1998) and can be summarized as follows:

(1) Identify all the local extrema, and then connect all the local maxima by a cubic spline line as the upper envelope.
(2) Repeat the procedure for the local minima to produce the lower envelope. The upper and lower envelopes should cover all the data between them.

The mean of upper and low envelope value is designated as m_1, and the difference between the signal $x(t)$ and m_1 is the first component

$$h_1 = x(t) - m_1 \tag{7}$$

(3) If h_1 is an IMF, then h_1 is the first component of *x*(*t*).
(4) If h_1 is not an IMF, h_1 is treated as the original signal and repeat steps (1–3); we get:

$$h_{11} = h_1 - m_{11}, \tag{8}$$

in which, m_{11} is the mean of upper and low envelope value of h_1. After repeated sifting, h_{1k} becomes an IMF, that is $h_{1k} = h_{1(k-1)} - m_{1k}$ then, it is designated $c_1 = h_{1k}$ as the first IMF component from the original data. c_1 should contain the finest scale or the shortest period component of the signal.

(5) Separate c1 from *x(t)*, we could get:

$$r_1 = x(t) - c_1 \tag{9}$$

r_1 is treated as the original data and repeat the above processes, the second IMF component c_2 of *x*(*t*) could be got.

(6) Let us repeat the process as described above for n times, then n-IMFs of signal *x* (*t*) could be got. Then,

$$\begin{gathered} r_1 - c_2 = r_2 \\ \vdots \\ r_{n-1} - c_n = r_n \end{gathered} \tag{10}$$

The decomposition process can be stopped when r_n becomes a monotonic function, from which no more IMF can be extracted.

By summing up (9) and (10), we finally obtain

$$x(t) = \sum_{j=1}^{n} c_j + r_n \tag{11}$$

The residue is the mean trend of *x*(*t*).

3.2 EEMD Algorithm

To alleviate the mode mixing effect of EMD, the EEMD was used. The EEMD decomposition algorithm of the original signal *x*(*t*) used in this work is summarized in the following steps (Wu and Huang 2009):

(1) Add a white noise *n*(*t*) with given amplitude β_k to the original signal *x*(*t*) to generate a new signal:

$$\mathrm{x_k(t) = x(t) + \beta_k n(t)} \tag{12}$$

(2) Use the EMD to decompose the generated signals $x_k(t)$ into N IMFs $IMF_{nk}(t), n = 1,\ldots,N$, where the nth IMF of the kth trial is $IMF_{nk}(t)$.
(3) Repeat steps (1) and (2) K times with different white noise series each time to obtain an ensemble of IMFs: $IMF_{nk}(t), k = 1, \ldots, K$
(4) Determine the ensemble mean of the K trials for each IMF as the final result:

$$\mathrm{IMF_n(t)} = \mathrm{lim_{k\to\infty}} \frac{1}{\mathrm{K}} \sum\nolimits_{\mathrm{k=1}}^{\mathrm{K}} \mathrm{IMF_{nk}(t)}, \quad \mathrm{n} = 1, \ldots, \mathrm{N} \tag{13}$$

4 Application

In order to highlight the efficiency of combined technique based mainly on two methods (EEMD and TFF) in early fault detection, we will implement it on simulations issued from a dynamic model of a gear transmission running under non-stationary conditions (variable load and speed). It was shown in the literature that the analysis of simulated vibration signals from gear models using Wigner Ville (Chaari et al. 2013) or spectrogram (Chaari et al. 2013; Bartelmus and Zimroz 2009) does not make it possible to detect teeth defects at an early stage. This is mainly caused by the fact that impacts induced by this localized defect are masked by the part of the signal with simultaneous amplitude and frequency modulation induced by speed and load variation. Let's consider a bevel gear transmission model is considered. The transmission is loaded with a torque having sawtooth shape with frequency fL = 5 Hz as presented in the Fig. 1a.

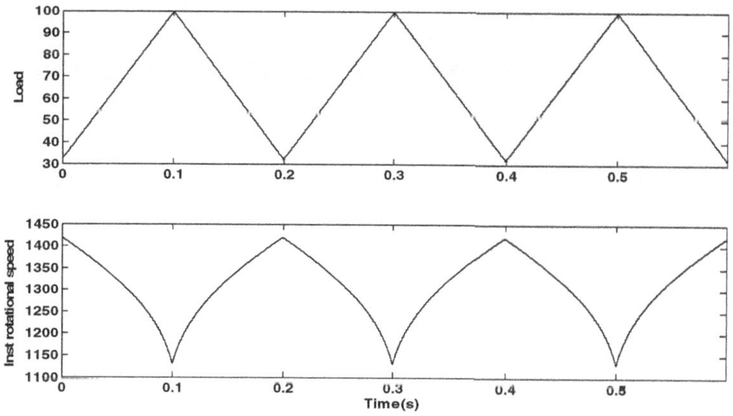

Fig. 1. (a) Evolution of the applied load, (b) Evolution of the instantaneous rotational speed

The variation of load leads to a fluctuation in the rotational speed (Fig. 1b) and a variation of the gear mesh frequency. The mean value of the motor rotational speed is n_r = 1320 rpm which corresponds to a mean gear mesh frequency f_{gm} = 308 Hz. The Inserted defect is a crack on one of pinion teeth (Z = 14 teeth) then the frequency default is 22 Hz which correspond to a period of 0.045 s. The sampling frequency is 30800 Hz for all signals. A crack in one pinion tooth is simulated by a periodic decrease in the gear mesh stiffness function corresponding to the mesh of the defected tooth.

In this work we propose to study acceleration signals for a load fluctuation of 50%, we have four different signals:

(a) *Gear without defect (0%),*
(b) *gear with a defect that the severity is 1%*
(c) *gear with a defect that the severity is 5%*
(d) *gear with a defect that the severity is 10%*

The acceleration signals for healthy gear and faulty gear for the early and advanced stage are given in Fig. 2.

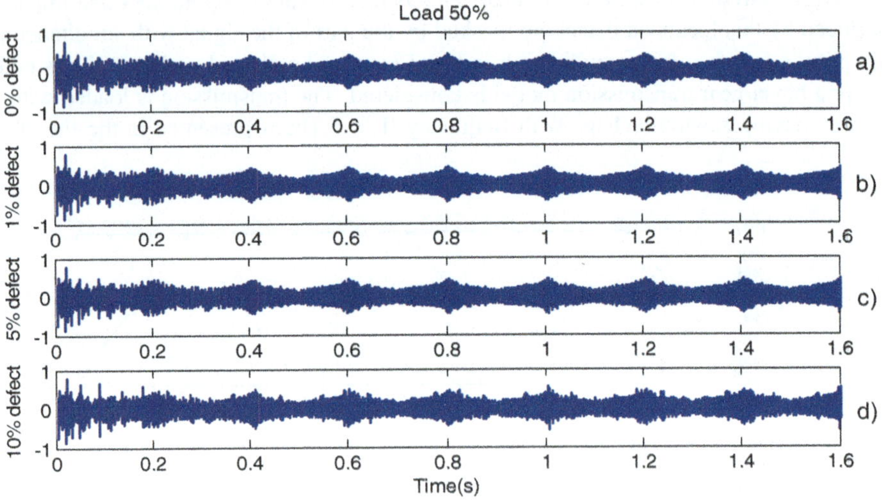

Fig. 2. Acceleration signals for 50% of the load. (a) Healthy gear, (b), (c) and (d) faulty gear.

From literature, the spectrum of a gear transmission running under constant loading conditions is dominated only by the gear mesh frequency and its harmonics with eventual sidebands induced by the presence of defects (Capdessus and Sidahmed 1992). For non-stationary conditions, family of sidebands will be noticed around the mesh frequency f_{gm} and its harmonics induced by the non uniformity of the gear mesh period (Fig. 3) and this can be thought to be a frequency modulation of the gear mesh stiffness.

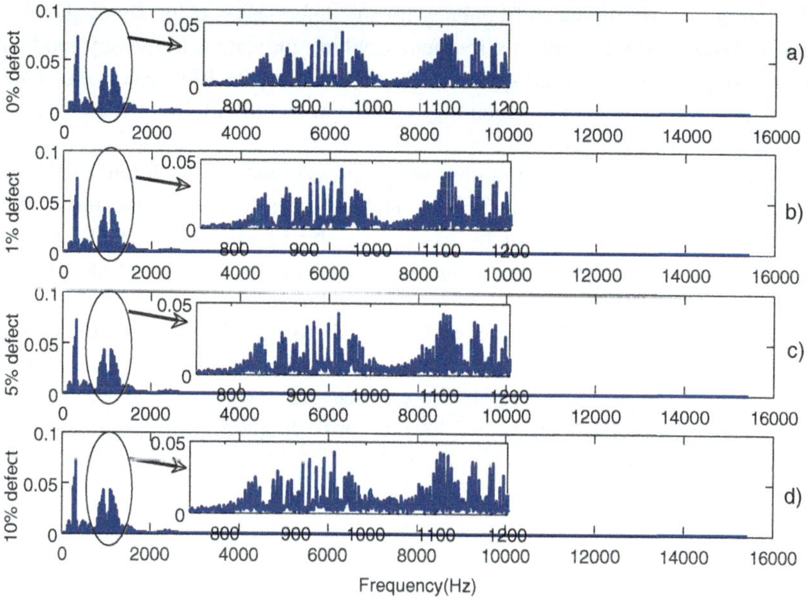

Fig. 3. Spectrum of the signals for 50% of load (a) healthy gear, (b), (c) and (d) Faulty gear.

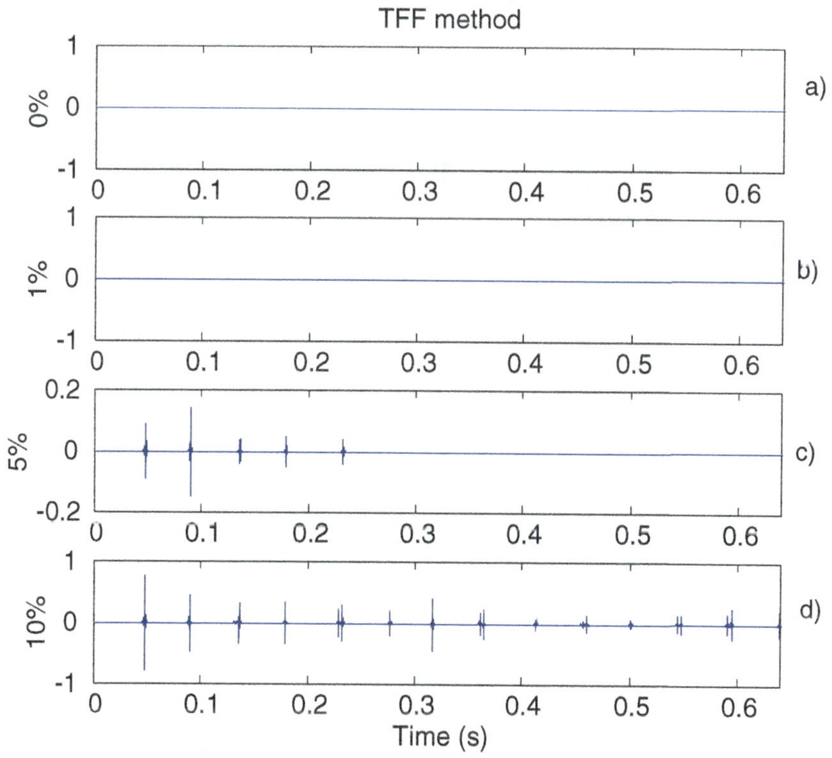

Fig. 4. TFF of signal for 50% of load (a) healthy gear, (b), (c) and (d) faulty gear

The zoom around the mesh frequency for the defect cases (Fig. 3b–d) shows many asymmetric sidebands around this frequency, which indicate a frequency modulation.

From the presented zoomed spectrum (Fig. 3a) for the healthy case, we can observe also the presence of sidebands which may cause confusion with the defected case when diagnosing the transmission.

In order to overcome this difficulty, we propose to use two different methods TFF and EEMD. Initially, we use the TFF to analyze the raw signals and the results are compared with the results given by using only the EEMD and then the results are compared with the results given by the combination of the two methods EEMD and TFF.

To compare the results, the acceleration signals for healthy gear and faulty gear for the early and advanced stage for 50% a variation of load are filtered by using TFF. The results obtained are given in Fig. 4. We can see that the method gives good results when the defect is very advanced and does not provide good information on the defect if it is early (5% or less than 5%).

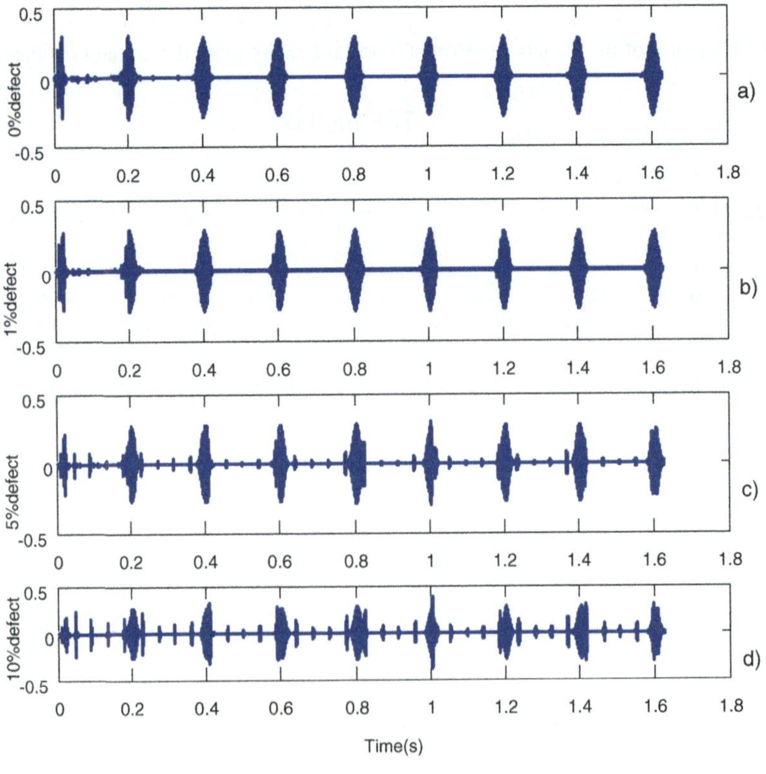

Fig. 5. IMF1 of signal for 50% of load (a) healthy gear, (b), (c) and (d) faulty gear

The same signals are also decomposed by using EEMD, Fig. 5 presents the first IMF of the four studied signals. From this figure we can see that for the case of faulty gear, we can observe clearly the variation of the load which can hide the impulsions due to the defect and precisely at early stage (5% defect), we can see also the position of impulses from 5% of severity which is the best result compared to results given by TFF, and we can see the impulses due to defect if the severity is greater than 10%. The period between to impulses is 0.045 s which is equivalent to the frequency defect.

EEMD allow us to detect the impulsions and eliminate the effect of the applied load when the defect is in advanced stage 10% and 5%. To eliminate completely the effect of the presence of a load, we propose to combine the two methods.

Figure 6 presents the first IMF of four studied signals after filtering by using TFF method. From this figure, we can see the impulses due to the defect if the severity is 5% but for a less severity, is 0% and 1% we see also some impulses which are the feature of the load.

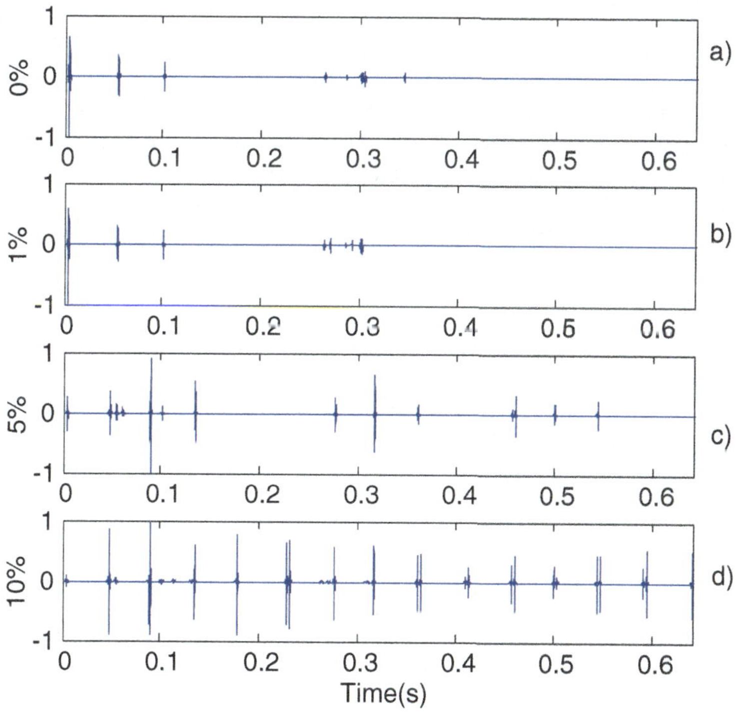

Fig. 6. IMF1 filtered using TFF (a) healthy gear, (b), (c) and (d) faulty gear

The TFF of these IMFs shows clearly the position of the impulses due to the fault. The periodicity of the defect can be clearly observed for 10% of defect better than 5% of the defect after using EEMD.

5 Conclusion

In this study, we have combined the EEMD and TFF methods to analyze non-stationary signals that give information about the variable conditions such as variable speed and load. The time-frequency filtering (TFF) method can be used to denoise signal and eliminate non stationary part as the effect of load. And the EEMD method achieves good modes separation. In this study, we have used the proprieties of the two methods to separate the fault effect from the load effect and to detect the fault masked by simultaneous variation of load. The EEMD method showed successful separation of the different modes that correspond to the variation of load and the effect of fault. We have used the EEMD method to decompose the signal in many modes and then we have used the TFF to detect the period of the impulses due to the fault and to eliminate completely the load's features from the modes by filtering.

References

McFadden, P.D.: Detecting fatigue cracks in gears by amplitude and phase demodulation of the meshing vibration. Trans. ASME J. Vib. Acoust. Stress Reliab. Design **108**, 165–170 (1986)

Mahgoun, H., Chaari, F., Felkaoui, A., Haddar, M.: Early detection of gear faults in variable load and local defect size using ensemble empirical mode decomposition (EEMD). In: Advances in Acoustics and Vibration Proceedings of the International Conference on Acoustics and Vibration (ICAV2016), Hammamet, Tunisia, 21–23 March 2016, pp. 13–23 (2016)

Capdessus, C., Sidahmed, M.: Analyse des vibrations d'un engrenage cepstre, corrélation, spectre. Traitement du Signal **8**(5), 365–371 (1992)

Cohen, L.: Time-frequency distributions a review. Proc. IEEE **77**(7), 941–981 (1989)

Forrester, B.D.: Use of Wigner Ville distribution in helicopter transmission fault detection in iProc of the Australian. In: A Symposium on Signal Processing and Applications, ASSPA 1989, Adelaide, Australia, 17–19 April 1989, pp. 77–82 (1989)

Staszewski, W.J.: Local tooth fault detection in gearboxes using a moving window procedure. Mech. Syst. Signal Process. **11**(3), 331–350 (1997)

Wang, W.J., Mcfadden, P.D.: Application of orthogonal wavelet to early gear damage detection. Mech. Syst. Signal Process. **9**(5), 497–507 (1997)

Mallat, S.G.: A Wavelet Tour of Signal Processing. Academic, San Diego (1998)

Liu, B., Riemenschneider, S., Xub, Y.: Gearbox fault diagnosis using empirical mode decomposition and Hilbert spectrum. Mech. Syst. Signal Process. **20**(3), 718–734 (2006)

Peng, Z.K., Chu, F.L.: Application of the wavelet transform in machine condition monitoring and fault diagnostics: a review with bibliography. Mech. Syst. Signal Process. **18**, 199–221 (2004)

Huang, N.E., Shen, Z., Long, S.R.: The empirical mode decomposition and the Hilbert spectrum for nonlinear and non-stationary time series analysis. In: Proceedings of the Royal Society of London Series, vol. 454, pp. 903–995 (1998)

Huang, N.E., Wu, M.L., Long, S.R.: A confidence limit for the empirical mode decomposition and Hilbert spectral analysis. Proc. Royal Soc. Lond. **459**, 2317–2345 (2003)

Rilling, G., Flandrin, P.: One or two frequencies? The empirical mode decomposition answers. IEEE Trans. Signal Process. **56**(1), 85–95 (2008)

Wu, Z., Huang, N.E.: Ensemble empirical mode decomposition: a noise-assisted data analysis method. Adv. Adapt. Data Anal. **1**(1), 1–41 (2009)

Mahgoun, H., Bekka, R.E., Felkaoui, A.: Gearbox fault diagnosis using ensemble empirical mode decomposition (EEMD) and residual signal. Mech. Ind. **13**(01), 33–44 (2012)

Wu, T.Y., Chung, Y.L.: Misalignment diagnosis of rotating machinery through vibration analysis via hybrid EEMD and EMD approach. Smart Mater, Struct. **18**(9) (2009)

Chaari, F., Abbes, M.S., Rueda, F.V., del Rincon, A.F., Haddar, M.: Analysis of planetary gear transmission in non-stationary operations. Front. Mech. Eng. **8**(1), 88–94 (2013)

Bartelmus, W., Zimroz, R.: Vibration condition monitoring of planetary gearbox under varying external load. Mech. Syst. Signal Process. **23**(1), 246–257 (2009)

Flandrin, P.: Time-frequency filtering based on spectrogram zeros (2015). https://hal-ens-lyon.archives-ouvertes.fr/ensl-01121522

Meignen, S., Oberlin, T., Depalle, P., Flandrin, P., McLaughlin, S.: Adaptive multimode signals reconstruction from time-frequency representations. Phil. Trans. R. Soc. (2016). http://rsta.royalsocietypublishing.org

Wu, T.Y., Chen, J.C., Wang, C.C.: Characterization of gear faults invariable rotating speed using Hilbert-Huang Transform and instantaneous dimensionless frequency normalization. Mech. Syst. Signal Process. **30**(1), 103–122 (2012)

Yang, W., Court, R., Tavner, C.J.: Crabtree Bivariate empirical mode decomposition and its contribution to wind turbine condition monitoring. J. Sound Vib. **330**(15), 3766–3782 (2011)

Mahgoun, H., Bekka, R.-E., Felkaoui, A.: Application of ensemble empirical mode Decomposition (EEMD) method for detection of localized faults in gear. In: IMPACT 2010, Djerba 22–24 mars (2010)

Comparison Between Hidden Markov Models and Artificial Neural Networks in the Classification of Bearing Defects

Miloud Sedira(✉), Ridha Ziani, and Ahmed Felkaoui

LMPA, Ferhat Abbas University, Setif, Algeria
miloudsedira@yahoo.fr, ziani_lmpa@yahoo.fr,
a_felkaoui@yahoo.fr

Abstract. In this paper a comparative study between two classification methods was presented, the first one belongs to the statistical domain in this case the Hidden Markov Models (HMM), the second is an Artificial Intelligence (AI) tool known as of Artificial Neural Networks (ANN), given their popularity in recent years and the interest shown by researchers in these methods, as to their performance and efficiency in the field of classification mainly. Indeed, the two classification tools were tested on data collected from vibratory signals on a test bench at the Bearing Data Center of Case Western Reserve University, and after being put in the appropriate form by an adequate signal processing and analysis to facilitate implementation. In this study, we have tried to identify the advantages and disadvantages of both tools in the field of classification of rotating machine defects, with the aim of accessing other work for the implementation of a classifier as effective as efficient. The results obtained are described as satisfactory and encouraging by their compatibility with those obtained by others implemented by other research but in other fields such as speech processing or image processing, which will give the character of originality to our work once completed.

Keywords: Bearing · Condition monitoring · Classification · HMM ANN · Signal processing

1 Introduction

The competitiveness of organizations and nations is directly affected by operational safety, the efficiency of maintenance costs and the availability of equipment conditional. Today, complex and developed production equipment requires highly sophisticated and costly maintenance strategies (Heng et al. 2009). The various demands and constraints imposed by productivity on the one hand, and the opportunities and solutions offered by the technology, notably computer technology and electronics, on the other, have fostered the continuous development of maintenance. From curative maintenance to immediate failure repair, to preventive maintenance based on scheduled maintenance operations, today arriving at predictive Maintenance (CBM) based on the anticipation and prediction of failures, resulting in a profit in time and cost and ensures benefit. Conditional Maintenance (CBM) consists of three phases: data acquisition,

A. Felkaoui et al. (Eds.): SIGPROMD'2017, ACM 12, pp. 68–78, 2019.
https://doi.org/10.1007/978-3-319-96181-1_6

data processing and decision-making. Diagnosis and prognosis are two important components of the CBM program (Jardine et al. 2006; Zhou et al. 2016; Kan et al. 2015). Vibration analysis has proven to be an effective means of diagnosing breakdowns. For decades, many methods have been proposed to extract and analyze the characteristics of the vibratory signal in order to perform the fault diagnosis (Wang et al. 2009). Considerable researches have been carried out on the development of various detection and diagnostic techniques based on vibratory data. These techniques can be classified mainly in the time domain, frequency domain, and time-frequency domain (Sanz et al. 2012). Increasingly sophisticated diagnostic methods have been used to determine the etiology of mechanical failure. Several diagnostic approaches are defined as, physical model, reliability model, machine learning model, and dependency model (Lee et al. 2014). The fault diagnosis is a procedure for mapping the information obtained in the measurement space and/or the characteristics in the characteristic space. This process is also called pattern recognition. Therefore, automatic pattern recognition is highly desirable. This can be achieved by classifying signals based on the information and/or characteristics extracted from the signals (Jardine et al. 2006).

Several defect classifiers based on characteristics extracted from vibratory signals, have been developed in this area. Recent approaches include K-nearest neighbors (KNN) the Gaussian mixture model (Wu et al. 2012), machine vector support (SVM) (Wu et al. 2012; Ziani et al. 2014; Sloin and Burshtein 2008; Zhu et al. 2009; Zhou et al. 2016; Li et al. 2015; Watanabe et al. 2013), and artificial neural network (Chen et al. 2014; Qiang et al. 2014; Wang et al. 2010; Sanz et al. 2012).

In this paper, a comparative study between two classifiers of different type is presented; the first one being an Artificial Intelligence (AI) tool, in this case Artificial Neural Networks (ANN), and the second is a probabilistic classifier consisting of Hidden Markov Models (HMM). These two tools have been used in the classification of ball bearing defects and then evaluate one with respect to the other, highlighting the advantages and disadvantages of one and the other. The following section presents the related works with our study.

2 Related Works

Several researchers around the world are working on the development of reliable and high-performance fault classifiers, with the aim of achieving diagnostic automation and subsequently the prognosis which is a primary objective of conditional maintenance (CBM). An important part of this research is focused on shape recognition tools of all types.

(a) Many researchers have opted for statistic tools. Thus, using the Gaussian mixing models, (Wang et al. 2009) succeeded in performing a classification of the bearing defects and their degree of gravity, without resorting to the extraction of the characteristics of the vibratory signal, estimating the delay and the dimensional integration Of the time series, the vibration signal is reconstructed in the phase space and then the Gaussian mixture model (GMM) is established for each type of defect signal in the phase space. In the same context of the static tools,

(Wang et al. 2009) used another model whose use in the field of mechanical diagnosis is very recent and remains relatively less explored, in this case the HMM, for the automatic diagnosis of bearing defects. The principle of this work is based on a given database established on the history of breakdowns listed in a catalog or dictionary. The HMM are then used to relate the data carried by a current signal with the established directory and to identify the defect by correspondence. The difficulty or inconvenience in this approach lies in the establishment of this catalog (Li et al. 2015). Also used the HMM for testing on simulation and real-life bearing fault diagnosis problem using the diversified gradient descent algorithm (DGD) to overcome the learning difficulties of HMM. This proposed formula does not require any particular form of objective function, it provides various estimates of parameters with different degrees of diversity, and it is obtained by dynamically adjusting the iterative procedure as a function of the gradient change of each parameter.

(b) Other researchers have used the tools of the artificial intelligence (AI) as (Ziani et al. 2014), carried out by the machine support vectors (SVM) on the same database, using Fisher criterion for the selection of the indicators extracts of the vibratory signal, in the same context (Wu et al. 2012), also used the SVM on the same database to automate the diagnosis using the characteristics extracted from the vibration signal by an entropy technique Of multi-scale permutation (MSE), for the recognition of rolling defects (Chen et al. 2014), used an ANN for the automatic diagnosis of rolling faults in the aforementioned database by characterizing the defects by a dependent characteristic vector (DFV) to designate symptom attributes. The characteristic vector is derived from the classification of the characteristics on the basis of the evaluation of the Euclidean distance. Whereas, with the same ANN but with multiple weights to form a probabilistic neural network (PNN) to diagnose rolling defects in the database of case western reserve university. Wang et al. (2010) have presented a hybrid approach of coupled pulse neural networks (PCNN) with the probability neural network (PNN) in order to perform an automatic diagnosis of a hydraulic generator. The first network is used for extracting characteristics of the vibratory signal in the time domain, while the second network is used to classify the defects of the generator in question. A multi-layer perceptron (MLP) was used by Sanz et al. (2012), for the classification of gear defects, the characteristics extracted by a wavelet packet transformation were used as a level of input data to the network. The second level was fed by the meshing data of the torsional stiffness of the gear-carrying shaft. A multi-stage algorithm was applied to supervise learning of the neural network.

(c) Another part of the researchers evaluated the classification tools by producing guides for users of the classification, like Miao et al. (2007), which presented a comparison of a statistical tool; Which is the HMM with an AI tool in this case the SVM, the advantages and disadvantages of one with respect to the other have been highlighted. This work is an important repertoire for researchers in this field. In the same way, we present in this paper, a comparison study between two classification tools, the first belongs to the field of statistics (HMM), the second to the domain of the AI (ANN). In the following section, we present the experimental equipment and the protocol of work procedure.

3 Apparatus and Experimentation

The vibratory signals that constitute the database, which supported the application of our study, were obtained from an experiment carried out at the Bearing Data Center of Case Western Reserve University (Ziani et al. 2014), and made available to unrestricted access for users on the website (http://csegroups.case.edu/bearingdatacenter/home). The reliability of this data base is justified by the many works published which can be cited (Heng et al. 2009; Raj and Murali 2013; Georgoulas et al. 2015; Li et al. 2014; Liu et al. 2015; Chebil et al. 2011; Rodriguez et al. 2013; Ziani et al. 2014; Watanabe et al. 2013; Wu et al. 2012; Chen et al. 2014; Tian et al. 2015; Qiang et al. 2014; Li et al. 2015).

Fig. 1. Experimental setup

The experimental setup shown in Fig. 1 consists of a 2 hp motor (left), a torque sensor/encoder (center), a dynamometer (right) and an electronic control not shown). Single-point defects were introduced on the inner ring, outer ring and ball using electro-discharge machining with diameters of 0.007 in., 0.014 in., 0.021 in. and 0.028 in. For more details, we invite the reader to consult the website mentioned above. Vibratory signals were collected for normal bearings, and bearings with various defects

(0.007 in., 0.014 in., 0.021 in. and 0.028 in.) (Fig. 2). Data were collected at 12,000 Hz for the fan end and at 48,000 Hz for the drive end bearings.

For our study, we only looked at the health of the fan bearing at the inner race with different health states, that is to say; the normal state, state with defect of 0.007″, state with defect of 0.014″, state with defect of 0.021″ and state with defect of 0.028″. The most important for this context is to freeze the same environment for the two cases, then evaluate the reaction of each of the two methods. The following section explains how the significant features have been obtained from the vibratory signals.

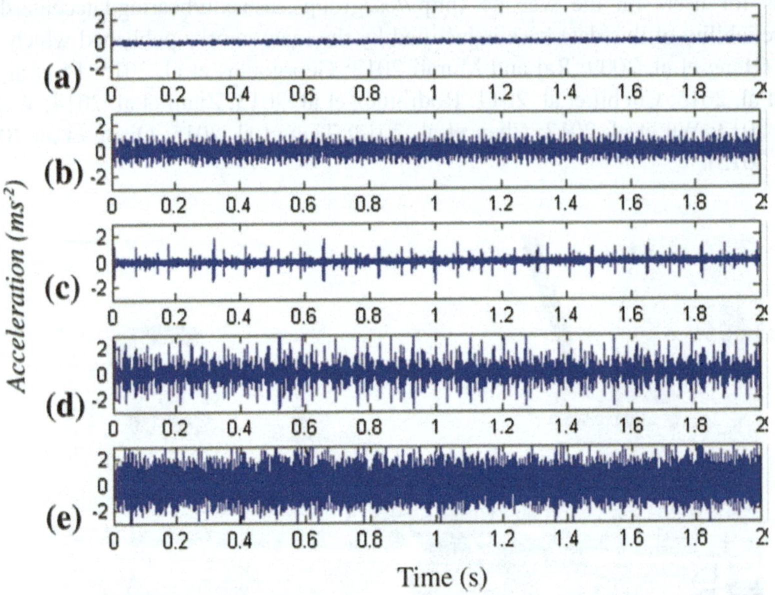

Fig. 2. Time domain signals acquired under 2 hp motor load for normal and faulty bearing with inner race fault. (a) Normal, (b) Fault diameter of 0.007 in., (c) Fault diameter of 0.014 in., (d) Fault diameter of 0.021 in., (e) Fault diameter of 0.028 in.

4 Data Analysis and Features Extraction

In fault diagnosis of rotating machines, the signature of the fault is mainly contained in the temporal vibration signals, which can be also represented in three forms: time domain, frequency domain and time-frequency domain. To ensure better characterization of the signal, it was preferred to extract the most significant features from each domain. In this context, we used the features established in the framework of a previous work by Ziani et al. (2014) knowing that it is the same experimental framework, the objective of our study is the comparison between two classifiers notwithstanding the diagnosis and finally this work is closely related to Ziani et al. (2014). Thus, in the time domain, seven statistical features were established: mean, peak factor, asymmetry, kurtosis and the central statistical moments of order 5 to 7 (Ziani et al. 2014). In the frequency domain, five (5) characteristics were calculated: the sum of the peaks of the power spectral density (PSD), calculated in 4 frequency bands centered on the fault frequency and its harmonics, and then in the total band. The third-order wavelet packet decomposition (WPD) in the time-frequency domain allowed us to calculate Kurtosis and energies for each coefficient; this allowed us to build the sets of learning and test data, which is illustrated in the following section.

5 Building Data Sets

Based on features extracted in the previous section, learning and test data sets were constructed for classification of bearing defects Table 1. The vibratory signals have been divided into 80 samples, for each rolling state or each class, there are 16 samples or sequences, divided into two groups of 8 samples each; One group reserved for learning and the other for the test. For a supervised classification, states or classes with their attributes have been designated as:

Classes or states: State 1, State 2, State 3, State 4, State 5, assigned attributes: Normal, default 0.007″, default 0.014″, default 0.021″ and default 0.028″.

Table 1. Data sets configuration

		Features								
		I_{11}	I_{12}	I_{13}	.	.	.	.	I_{140}	
Classes	C_1	.	.	.	.	.	.	.		Learning data
	C_2	.	.	.	.	.	.	.		
	.	.	.	.	.	.	.	.		
	C_{40}	.	.	.	.	.	.	.	I_{4040}	
	.	.	.	.	.	.	.	.		Test data
	C_{80}	.	.	.	.	.	.	.	I_{8040}	

The classes in question with their attributes will serve as input data for the learning of the two classifiers, this is what will be illustrated in the next section.

6 Application of Hidden Markov Models and Artificial Neural Networks in the Classification of Bearing Defects

6.1 Theoretical Background

6.1.1 HMM

In recent years, we have witnessed a growing interest in HMM, given their powerful ability to model time-varying signals in many real-world applications (Rabiner 1989; Kan et al. 2015; Li et al. 2015). An HMM is a statistical approach based on the Markov chain principle to model signals that evolve through a finite number of states. A Markov chain is a sequence of states where each depends only on the event that immediately precedes it (Kan et al. 2015). For more theoretical details see Rabiner (1989). An HMM is a doubly stochastic process, with an underlying process hidden from finite states, which associates itself with the observation process (Miao et al. 2007; Zhu et al. 2009; Lee et al. 2014). An HMM is characterized by: An initial state distribution vector π, a transition matrix A, an emission matrix B (1). An HMM is designated by:

$$\lambda = (A, B, \pi) \tag{1}$$

6.1.2 ANN

ANN are a pattern recognition tool belonging to the concept of AI. The basic idea of this concept has been extracted from the functioning of the human brain. They have excellent learning and generalization skills nevertheless, they can deviate from complicated or imprecise data (Kan et al. 2015) because they are generally used for the classification of static inputs without sequential processing. They can also create their own representation of the received information. These exceptional features make ANN a powerful tool for modeling data (Kan et al. 2015). The minimization of the objective function, also called the cost function, is the main learning problem encountered in the ANN classification. One of the most used learning algorithms is back propagation (Pacheco et al. 2016; Sanz et al. 2012).

6.2 Application

6.2.1 HMM

Typically, an HMM is assigned to each class and its parameters are estimated from a learning database using the maximum likelihood (ML) method. The recognition of an observed sequence which represents an unknown class can then be done by estimating the parameters of the HMM related to it, hence the unobservable state sequence, can be estimated by the observation sequence (Wang et al. 2010; Rabiner 1989; Zhou et al. 2016). The basis established in Sect. 4. contains 80 samples of 40 features for the five states defined above, the learning was done on 3 steps, the first with 10 features, the second with 20 features, and the third with all the Features i.e. 40 features. Indeed for each class or state an HMM is computed, characterized by its, log-likelihood (LL), transition matrix and emission matrix. The results are summarized in Table 2.

$$LL = \log P(O|\lambda). \tag{2}$$

λ is the HMM, O is the observation, LL is the log likelihood of HMM, P is the probability

Table 2. HMM Recognition performance

HMM	λ_1	λ_2	λ_3	λ_4	λ_5
State	1	2	3	4	5
Recognition performance (%)					
10 features	95.75	97.66	98.28	97.08	96.93
20 features	98.09	98.07	96.14	98.07	96.37
40 features	98.23	94.47	93.96	93.62	98.23

6.2.2 ANN

A multilayer perceptron neural network (MLP) has been used in the present work to identify the different classes based on the severity of the defects. ML consist of an input layer of source nodes, one or more hidden layers of compute nodes and an output layer (Sanz et al. 2012). The same context observed for HMMs in Sect. 6.1.2 was applied for ANNs; the results of the classification are illustrated in Table 3.

Table 3. ANN Recognition performance

	ANN based classifier
	Test success (%)
10 features	99.00
20 features	97.00
40 features	95.50

7 Comparison Between HMM and ANN Based Classifiers

The comparison is made first based on the results generated by each of the classifiers, secondly on the context of data manipulation and the computational computation with the difficulties encountered. On this aspect, the ANN offers more facilities for modeling and generalization in relation to HMM, but the latter remain robust to the variation of data in quantity and quality (nature and type of values). The results of this comparison are shown in Table 4.

Table 4. Performance comparison between HMM and ANN based classifiers

	HMM based classifier	ANN based classifier
	Test success mean (%)	Test success (%)
10 features	93.34	99.00
20 features	97.50	97.00
40 features	89.61	95.50

8 Conclusion

A comparison between the hidden Markov models HMM and the artificial neural networks ANN is presented in this paper, for classification of ball bearing defects. This is a study of pattern recognition techniques for designing an efficient and robust classification system. Based on the results of the analysis, we found that the performance of the ANN decreases with the increase in the number of input data, while the HMM remain robust on this aspect. All the ANN are easier for computational modeling and manipulation in relation to HMM. This leads us to explore them deeply in future work.

Acknowledgements. This work is partially supported by the laboratory of applied precision mechanics (LMPA), Ferhat Abbas University, Setif, Algeria. The authors thank Professor K. LOPARO of case western university for providing the data. Also, the authors gratefully acknowledge the reviewers for their valuable comments and valuable suggestions, which greatly contributed to the improved presentation of this work.

References

Jardine, A.K.S., Lin, D., Banjevic, D.: A review on machinery diagnostics and prognostics implementing condition-based maintenance. Mech. Syst. Signal Process. **20**, 1483–1510 (2006). https://doi.org/10.1016/j.ymssp.2005.09.012

Heng, A., Zhang, S., Tan, A.C.C., Mathew, J.: Rotating machinery prognostics: state of the art, challenges and opportunities. Mech. Syst. Signal Process. **23**, 724–739 (2009). https://doi.org/10.1016/j.ymssp.2008.06.009

Sloin, A., Burshtein, D.: Support vector machine training for improved hidden markov modeling. IEEE Trans. Signal Process. **56**(1), 172–188 (2008)

Wang, C., Zhou, J., Wang, Y., Huang, Z., Kou, P., Zhang, Y.: Hybrid neural network based fault diagnosis of rotating machinery. In: 2010 3rd International Congress on Image and Signal Processing (CISP2010). IEEE (2010). 978-1-4244-6516-3/10/©2010

Raj, A.S., Murali, N.: Morlet wavelet UDWT denoising and EMD based bearing fault diagnosis. Electronics **17**(1), 1–8 (2013)

Pacheco, F., de Oliveira, J.V., Sánchez, R.-V., Cerrada, M., Cabrera, D., Li, C., Zurita, G., Artés, M.: A statistical comparison of neuro classifiers and feature selection methods for gearbox fault diagnosis under realistic conditions. Neurocomputing **194**, 192–206 (2016). https://doi.org/10.1016/j.neucom.2016.02.028

Georgoulas, G., Karvelis, P., Loutas, T., Stylios, C.D.: Rolling element bearings diagnostics using the symbolic aggregate approximation. Mech. Syst. Signal Process. **60–61**, 229–242 (2015). https://doi.org/10.1016/j.ymssp.2015.01.033

Wang, G.F., Li, Y.B., Luo, Z.G.: Fault classification of rolling bearing based on reconstructed phase space and Gaussian mixture model. J. Sound Vib. **323**, 1077–1089 (2009). https://doi.org/10.1016/j.jsv.2009.01.003

Li, H., Zhao, J., Zhang, X., Teng, H., Yang, R., Hao, L.: Bearing fault diagnosis method using envelope analysis and euclidean distance. Indones. J. Electr. Eng. **12**(3), 1887–1894 (2014). https://doi.org/10.11591/telkomnika.v12i3.4186

Zhou, H., Chen, J., Dong, G., Wang, R.: Detection and diagnosis of bearing faults using shift-invariant dictionary learning and hidden Markov model. Mech. Syst. Signal Process. **72–73**, 65–79 (2016). https://doi.org/10.1016/j.ymssp.2015.11.022

Liu, H., Wang, X., Lu, C.: Rolling bearing fault diagnosis based on LCD–TEO and multifractal detrended fluctuation analysis. Mech. Syst. Signal Process. **60–61**, 273–288 (2015). https://doi.org/10.1016/j.ymssp.2015.02.002

Chebil, J., Hrairi, M., Abushikhah, N.: Signal analysis of vibration measurements for condition monitoring of bearings. Aust. J. Basic Appl. Sci. **5**(1), 70–78 (2011)

Lee, J., Wu, F., Zhao, W., Ghaffari, M., Liao, L., Siegel, D.: Prognostics and health management design for rotary machinery systems – reviews, methodology and applications. Mech. Syst. Signal Process. **42**, 314–334 (2014). https://doi.org/10.1016/j.ymssp.2013.06.004

Sanz, J., Perera, R., Huerta, C.: Gear dynamics monitoring using discrete wavelet transformation and multi-layer perceptron neural networks. Appl. Soft Comput. **12**, 2867–2878 (2012). https://doi.org/10.1016/j.asoc.2012.04.003

Zhu, K., San Wong, Y., Hong, G.S.: Multi-category micro-milling tool wear monitoring with continuous hidden Markov models. Mech. Syst. Signal Process. **23**, 547–560 (2009). https://doi.org/10.1016/j.ymssp.2008.04.010

Rabiner, L.R.: A tutorial on hidden Markov models and selection. Proc. IEEE **77**(2), 257–286 (1989)

Kan, M.S., Tan, A.C., Mathew, J.: A review on prognostic techniques for non-stationary and non-linear rotating systems. Mech. Syst. Signal Process. **62–63**, 1–20 (2015). https://doi.org/10.1016/j.ymssp.2015.02.016

Miao, Q., Huang, H.-Z., Fan, X.: A comparison study of support vector machines and hidden Markov models in machinery condition monitoring. J. Mech. Sci. Technol. **21**, 607–615 (2007)

Rodrıguez, P.H., Alonso, J.B., Ferrer, M.A., Travieso, C.M.: Application of the Teager-Kaiser energy operator in bearing fault diagnosis. ISA Trans. **52**, 278–284 (2013). https://doi.org/10.1016/j.isatra.2012.12.006

Ziani, R., Felkaoui, A., Zegadi, R.: Bearing fault diagnosis using multiclass support vector machines with binary particle swarm optimization and regularized Fisher's criterion. J. Intell. Manuf. **28**(2), 405–417 (2014). https://doi.org/10.1007/s10845-014-0987-3

Watanabe, S., Nakamura, A., Juang, B.-H.F.: Structural bayesian linear regression for hidden Markov models. J. Signal Process. Syst. **74**(3), 341–358 (2013). https://doi.org/10.1007/s11265-013-0785-8

Wu, S.-D., Wu, P.-H., Wu, C.-W., Ding, J.-J., Wang, C.-C.: Bearing fault diagnosis based on multiscale permutation entropy and support vector machine. Entropy **14**, 1343–1356 (2012). https://doi.org/10.3390/e14081343

Chen, X., Zhou, J., Xiao, J., Zhang, X., Xiao, H., Zhu, W., Fu, W.: Fault diagnosis based on dependent feature vector and probability neural network for rolling element bearings. Appl. Math. Comput. **247**, 835–847 (2014). https://doi.org/10.1016/j.amc.2014.09.062

Tian, Y., Ma, J., Lu, C., Wang, Z.: Rolling bearing fault diagnosis under variable conditions using LMD-SVD and extreme learning machine. Mech. Mach. Theory **90**, 175–186 (2015). https://doi.org/10.1016/j.mechmachtheory.2015.03.014

Qiang, Y., Chen, L., Hua, L., Gu, J., Ding, L., Liu, Y.: Research on the classification for faults of rolling bearing based on multi-weights neural network. Int. J. Smart Sens. Intell. Syst. **7**(3), 1004–1023 (2014)

Li, Z., Fang, H., Huang, M.: Diversified learning for continuous hidden Markov models with application to fault diagnosis. Expert Syst. Appl. **42**, 9165–9173 (2015). https://doi.org/10.1016/j.eswa.2015.08.027

On-line Adaptive Scaling Parameter in Active Disturbance Rejection Controller

Maroua Haddar[1(✉)], S. Caglar Baslamisli[2], Fakher Chaari[1], and Mohamed Haddar[1]

[1] Mechanics, Modeling and Production Laboratory (LA2MP), Mechanic Departement, National Engineering School of Sfax (ENIS), BP 1173, 3038 Sfax, Tunisia
HADDAR.marwaGEM1@gmail.com, Fakher.chaari@gmail.com, Mohamed.haddar@enis.rnu.tn

[2] Department of Mechanical Engineering, Hacettepe University, Beytepe, 06800 Ankara, Turkey
Caglar.baslamisli@gmail.com

Abstract. Active Disturbance Rejection Controller (ADRC) is considered one of the most famous model free controllers in the industry. This introduced scheme of control, do not require the exact modeling of the system equations and used to reject online any types of perturbations. However, the drawback of this tool is the hard task of tuning multi-parameters and takes a long time to achieve performances requirements. In this contribution, an optimization of a scaling parameter which has an important effect in the dynamic behavior of controlled system. There has been some research concentrate in estimate the parameters uncertainties from input and output signals of the body mass in vehicle system. This kind of estimation is based on differential algebra which is known by its simplicity of implementation, fast and robust to noise marring any measured signals. Furthermore, the combination of this algebraic methodology with aforementioned control low is easy. For the purpose of improving the effectiveness of ADRC controller, this paper use to predict this unknown variation and it was incorporated in the equation of control. Using this time varying parameter instead of an empirical one, simulations results show an amelioration of the energy consumption and an increase of the ride comfort.

Keywords: Model free control · On-line estimation · ADRC
Sprung mass variation

1 Introduction

Recently, controllers which are independent on the mathematical model spread in many filed such as Model Free Control (Fliess and Join 2013) and Active Disturbance Rejection Control (ADRC) (Han 2009).

The ADRC technique used to estimate endogenous and exogenous perturbations with a state observer. This estimated states can be injected in the equation of controller and cancel all unknowns' phenomena. Unfortunately, for obtaining the optimal control requires a lot of time and needs many essays. The main difficulty of the calibration task

A. Felkaoui et al. (Eds.): SIGPROMD'2017, ACM 12, pp. 79–86, 2019.
https://doi.org/10.1007/978-3-319-96181-1_7

is to define properly the parameter that affect the denominator of controller. The range of variation of this scaling parameter is considered un-known and differ from an operator to another one.

This scheme of control was applied to control the passive quarter car system (Hasbullah et al. 2015; Li et al. 2016). Researchers have developed this method with a constant scaling parameter chosen by the operator and are approximate it to the inverse of sprung mass. However, our aim is to find a global approach by changing controller parameter in order to follow the sprung mass variation.

Recently, Alvarez-Sanchez (2013) provided an identification scheme framework to estimate the sprung mass variation based on algebraic rules of Fliess and Sira-Ramírez (2003). Using this approach, the range variation characteristic of body mass can be deduced based on the measured information of sprung mass displacement, un-sprung mass displacement and actuator force. Based on these mentioned principles, ADRC control using real-time identification for time-varying mass is proposed in this paper. A combination between online real-time identification and ADRC control can be used in order to escape the time-consuming in founding the optimal control.

The organization of the paper as follows. Section 2 describes the motion equations of quarter car model. Section 3 gives a simple description of estimator approach. The online estimation of sprung mass is discussed in Sect. 4. Results are presented in Sect. 5 and a conclusion is summarized in the last section.

2 Description of the System and Road Input

The motion equations of passive quarter car system are given below:

$$m_s \ddot{z}_s + d_s(\dot{z}_s - \dot{z}_u) + k_s(z_s - z_u) = F_A \tag{1}$$

$$m_u \ddot{z}_u - d_s(\dot{z}_s - \dot{z}_u) - k_s(z_s - z_u) + k_t(z_u - z_r) = -F_A \tag{2}$$

m_s is the sprung mass which represents the body of the car. m_u is the un-sprung mass. k_s is the suspension stiffness. d_s Represents suspension damping and k_t is the tire stiffness. The actuator force is denoted by F_A (Fig. 1).

A random road excitation is characterized by a constant of roughness that is given by:

$$\dot{z}_r(t) = -2\pi V n_0 z_r(t) + 2\pi\sqrt{G_0 V\, w(t)} \tag{3}$$

where $z_r(t)$ is the random road displacement, V is the vehicle speed, the reference spatial frequency n_0, G_0 is the road roughness coefficient and the white noise signal is given by $w(t)$.

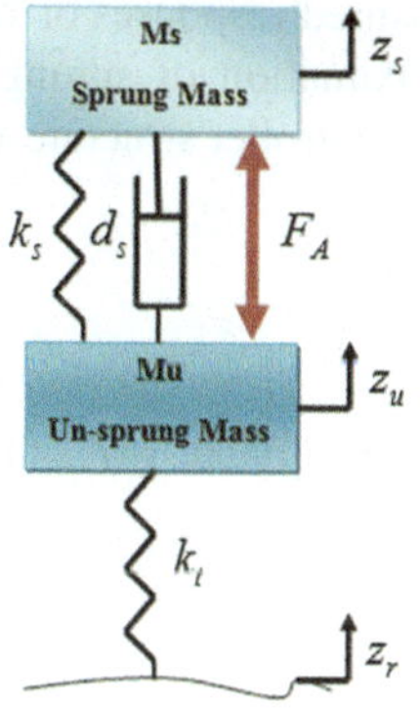

Fig. 1. Active quarter car model

Table 1. Parameters of suspension system

Parameters	Description	Value
m_s	Sprung mass	285 kg
m_u	Un-sprung mass	41 kg
k_s	Spring stiffness	17900 N/m
d_s	Damping constant	535 N/(m/s)
k_t	Tire stiffness	19125 N/m

3 Controller Design

The basic idea of ADRC is using an extended observer ESO without a priori information of the system. The attractiveness of this tool of control results from its capability to estimate on-line and reject unknown perturbations. Significant performances are achieved in the works of Hasbullah et al. (2015) and Pan et al. (2015).

In order to improve the response of suspension system described in Eqs. (1) and (2), the ADRC strategy used to reformulate these equations as,

$$\ddot{z}_s = h(t, \dot{z}_s, z_s) + \beta F_A + z_r \tag{4}$$

From practical point of view, the knowledge of h (.) and β is not straightforward. A non trivial approximation is used in other studies. Where, the constant β is chosen empirically and is approximated by $1/m_s$.

In real conditions, significant uncertainties of sprung mass can affect the behavior of this proposed controller. Estimation of this calibrating parameter can be a source of upgradability in energy consumption.

4 "β" On-Line Adaptation

An algebraic estimator used to find the car body mass of the quarter car system based on algebraic identification methods (for more details see Fliess and Sira-Ramirez 2003).

$$m_s(t)(2\int\int z_s - 4\int tz_s + t^2 z_s)dt + d_s(2\iint tz_s + \int t^2 z_s - 2\iint tz_u - \int t^2 z_u)dt + k_s(\iint t^2 z_s - \iint t^2 z_u)dt = \iint t^2 F_A dt \tag{5}$$

From the equation, it can be seen that only from the measured responses of vertical displacements and actuator force; we can obtain an approximation of sprung mass variation. This On-line estimation can be easily added to the controller structure where,

$$\hat{\beta}(t) = \frac{1}{\hat{m}_s(t)} \tag{6}$$

The structure of classical controller is changed with the variable $\hat{\beta}(t)$, the new design is given in this equations:

(1) the system's state:

$$\begin{cases} \dot{x}_1 = x_2 \\ \dot{x}_2 = x_3 + \hat{\beta}(t)F_A \\ \dot{x}_3 = \dot{f}(t, x_1, x_2, w) \end{cases} \tag{7}$$

where $x_1 = z_s, x_2 = \dot{z}_s$ and x_3 represents all the un-known perturbations.

(2) The linear Luenberger observer:

$$\begin{cases} \dot{z}_1 = z_2 - L_{01}\hat{e} \\ \dot{z}_2 = z_3 - L_{02}\hat{e} + \hat{\beta}(t)F_A \\ \dot{z}_3 = -L_{03}\hat{e} \end{cases} \tag{8}$$

The observer gains $[L_{01}, L_{02}, L_{03}]$ depends to the location the desired close loop poles. The approximated error is $\hat{e} = z_s - z_1$.

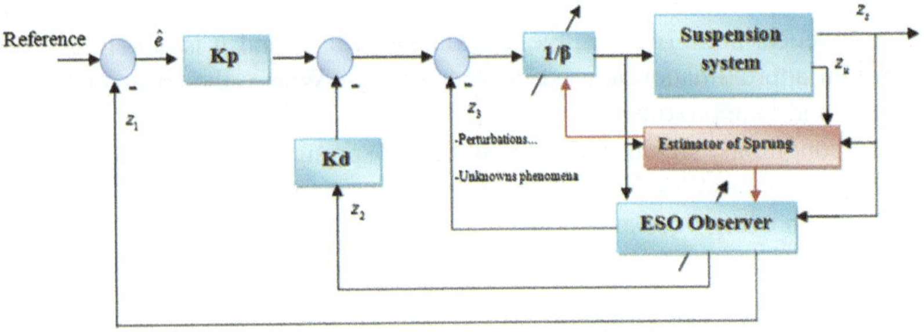

Fig. 2. Block diagram of the proposed control

The principle of combination between online real-time identification and ADRC control is described in Fig. 2.

The gains K_p and K_d are respectively, the proportional gain and the derivative gain, which are needed to implement the proportional-derivative (PD) feedback controller.

5 Results of Simulation

In the simulation algorithm a solver set to ODE5 and fixed integration step of 1 ms were used. The parameters of Suspension system are given by Table 1.

Figure 3 depicts the real value of mass and the identified masse $\hat{m}_i$ when using the "Eq. (5)". In reality, the on-line estimation process is characterized with few irregularities in the beginning. In order to eliminate this perturbation caused by the singularities; the implementation of identifier process is carried out at t > 0 s. It is observable that the estimation process is achieved after a short time t = 0.004 s. (In the rest of simulation, in the beginning β is chosen constant and independents of load variation. After that the estimation and the adaptive scheme start at t = 10 s).

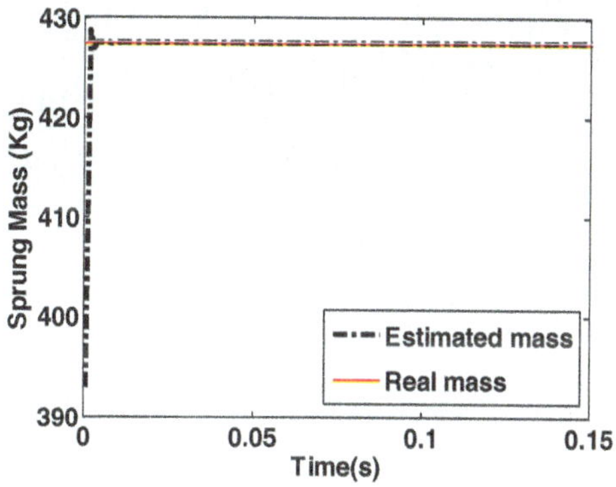

Fig. 3. On-line estimation of Sprung mass

Figure 4 shows the suspension deflection with the effect of sprung mass variation $m_s = +50\%\ m_{s\,initiale}$. The line at t = 10 s represents the instant where the identification started and applied to the controller equation. Before estimation of the constant from measured signals, the β is chosen empirically and it is not calibrated when the

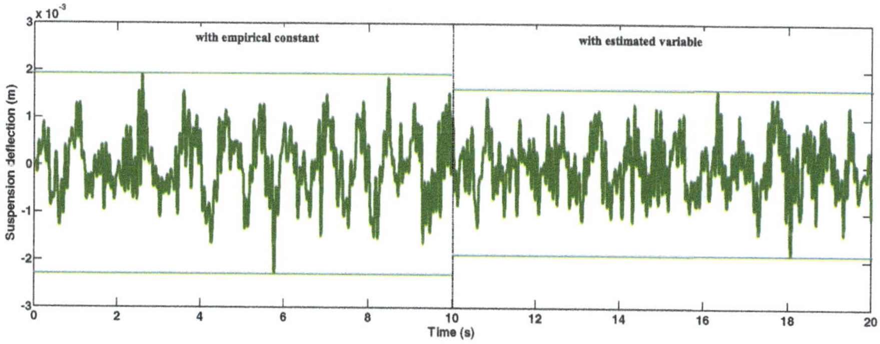

Fig. 4. Suspension deflection

body mass change. The ADRC controller with estimated $\hat{\beta}$ can produce more best tracking position than the ADRC without constant adaptation.

For quarter car control, the ride comfort is related to the body acceleration. In Fig. 5, we can see that the designed system conserve the best isolation of disturbance with a slight attenuation, for the reason that ADRC with adaptive $\hat{\beta}$ has the least RMS value among the sprung mass variation Fig. 6.

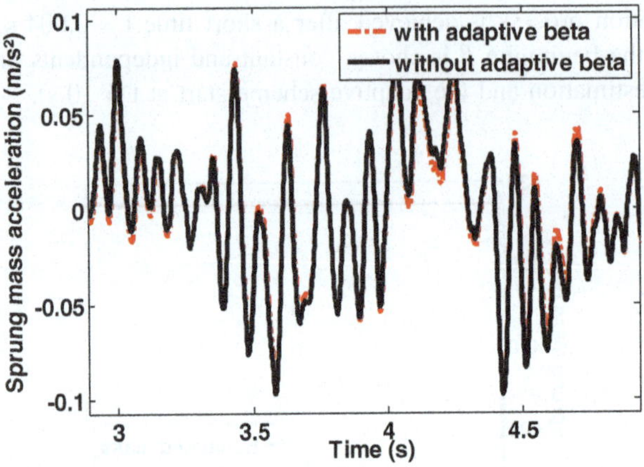

Fig. 5. Sprung mass acceleration variation

In Fig. 7 we observe a reduction of actuator force when we use the adaptive controller. According to this numerical results, the estimated $\hat{\beta}$ has an important role in the dynamic behavior of the suspension system. It is quite possible to obtain better sprung mass acceleration attenuation and the best tracking of reference trajectory for the ADRC controller with the lowest power demand.

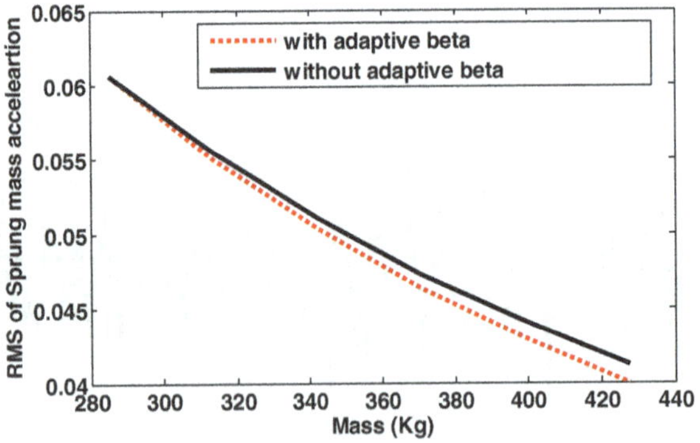

Fig. 6. RMS of Sprung mass acceleration

Furthermore, the Integral of the Square of the Error (ISE) performance was calculated with both of controller and with different masses. The results are depicted in Fig. 8.

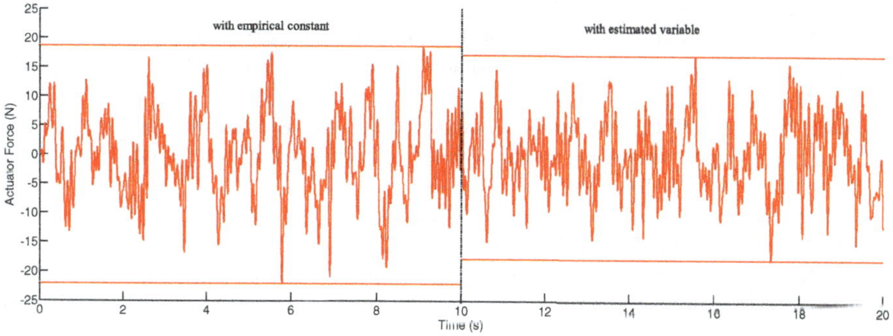

Fig. 7. The power demand of actuator

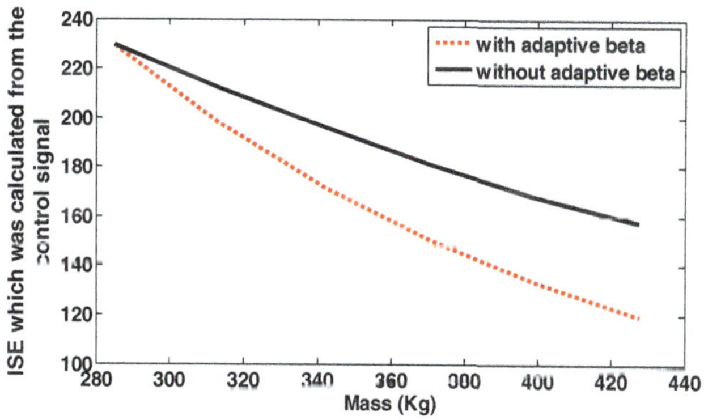

Fig. 8. ISE criteria was calculated from the power demand

6 Conclusion

The aim of this paper is to propose an adaptive ADRC controller using an On-line identification of Sprung mass which is intended to extended time-varying observer. This method can overcome time-consuming which are induced by nontrivial calibration. At the same time, this adaptive control provides high performances under sprung mass uncertainty. Lastly, this method applied to quarter car system permits to getting the best ride comfort and the best tracking with the minimum of power demand.

References

Alvarez-Sánchez, E.: A quarter-car suspension system: car body mass estimator and sliding mode control. Procedia Technol. **7**, 208–214 (2013)

Fliess, M., Sira-Ramírez, H.: An algebraic framework for linear identification. ESAIM: Control Optim. Calc. Var. **9**, 151–168 (2003)

Fliess, M., Join, C.: Model-free control. Int. J. Control **86**(12), 2228–2252 (2013)

Han, J.: From PID to active disturbance rejection control. IEEE Trans. Industr. Electron. **56**(3), 900–906 (2009)

Hasbullah, F., Faris, W.F., Darsivan, F.J., Abdelrahman, M.: Ride comfort performance of a vehicle using active suspension system with active disturbance rejection control. Int. J. Veh. Noise Vib. **11**(1), 78–101 (2015)

Li, P., James, L., Kie, C.C.: Experimental investigation of active disturbance rejection control for vehicle suspension design. Int. J. Theor. Appl. Mech. **1**, 89–96 (2016)

Pan, H., Sun, W., Gao, H., Hayat, T., Alsaadi, F.: Nonlinear tracking control based on extended state observer for vehicle active suspensions with performance constraints. Mechatronics **30**, 363–370 (2015)

Modal Analysis of the Clutch Single Spur Gear Stage System with Eccentricity Defect

Ahmed Ghorbel(✉), Moez Abdennadher, Lassâad Walha, Becem Zghal, and Mohamed Haddar

Laboratory of Mechanics, Modeling and Production, National School of Engineering of Sfax, University of Sfax, Sfax, Tunisia
gh.ahmed.enis@gmail.com

Abstract. Gears are an important element in a variety of industrial applications. An unexpected failure of the gear may cause significant economic losses. For that reason, fault diagnosis in gears has been the subject of intensive research. Modal analysis can be used in the fault detection of rotating machinery. It can provide natural frequencies and vibration modes which are essential information to learn about most of dynamic characteristics of the combined system. In order to investigate the dynamic behavior of a coupled clutch-gear transmission system in the presence of gear defect, a general dynamic model is developed and a numerical modal analysis technique is achieved. Several types of gear defects that can be found in the literature. In this paper, a gear eccentricity defect is introduced in the model to study their influence on the modal properties. The distributions of modal kinetic and strain energies are presented in the case without and with defect on the geared system, and a comparative study is conducted.

Keywords: Modal analysis · Clutch · Gear · Eccentricity defect
Kinetic and strain energies distribution

1 Introduction

Gears are amongst the frequently encountered components to be found in rotating machinery used in various applications. The literature is rich in theoretical and experimental works achieved on the gearings.

For gear transmissions, the two most commonly encountered manufacturing defects are the eccentricity error and the profile error. The eccentricities of the gears can introduce particularly noisy mounting configurations. Many works of research (Fakhfakh et al. 2006; Walha et al. 2009) include the different types of defects that may be affecting the gearings. Indeed, the researchers are interested on the defects gearings to be able to analyze the dynamic behavior of the transmission in the presence of these defects. Numerous research works have treated an automotive clutch. Gaillard and Singh (2000) proposed five minimal clutch models and studied the energy dissipation for each of them. Duan and Singh (2006) developed a torsional model for an automotive clutch by converting torque and dry friction disc. In another work, Walha et al. (2011) treated the effects of the eccentricity defect on the nonlinear dynamic behavior

A. Felkaoui et al. (Eds.): SIGPROMD’2017, ACM 12, pp. 87–95, 2019.
https://doi.org/10.1007/978-3-319-96181-1_8

of the mechanical clutch-helical two stage gear. More coupled models were also studied in the literature (Caruntu et al. 2016; Ghorbel et al. 2017a, b). All previous investigations do not have to study the modal proprieties and the vibration modes for coupled system.

In this paper, we propose a new model of coupled clutch-single stage spur gear system that contains twelve degrees of freedom. A manufacturing defect in the gear system is included to study their effect on the energies distribution. The natural frequencies and vibration modes are investigated. For each vibration mode, the modal kinetic and strain energies are also discussed.

2 Numerical Model

2.1 Combined Clutch-Transmission Model

A simplified transverse-torsional combined model with twelve degrees of freedom (DOF) was used to in this research work and as the focus of this study is to investigate the distribution of the kinetic and strain energies.

Figure 1 shows the global model which consisting two main subsystems; dry clutch and the gear transmission. The displacements of the bearing are modeled by linear stiffness k_{xi} and k_{yi} in x and y the directions (i = 1, 2 or 3). The connected shafts have a torsional stiffness K_1, K_2 and K_3. The connection between the flywheel and the clutch plate is modeled by a torsional stiffness k_f. In Fig. 1, T_e represents the engine torque and T_d is the load torque.

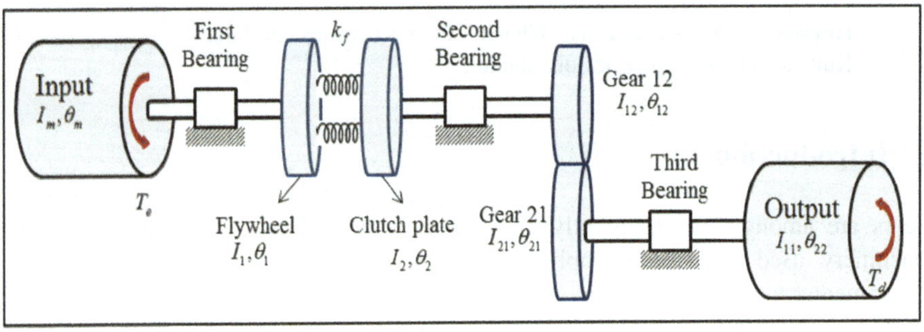

Fig. 1. Dynamic model of the clutch–spur gear stage.

I_m and I_{22} are respectively the inertias of the input and output. I_1 represents the combined torsional inertia of a flywheel. I_2 is the inertia of the friction disc and pressure plate. I_{12} and I_{21} represent the inertias of the gear-12 and gear-21.

2.2 Equation of Motions

The resulting equations of motion in matrix form are defined by:

$$[M]\{\ddot{q}\} + [K(t)]\{q\} = \{F\} \tag{1}$$

where M and C are respectively the mass and the damping matrices. $K(t)$ is the global stiffness matrix and F is the linear vector force.

The mass matrix can be expressed by

$$[M] = diag(m_{b1}, m_{b1}, m_{b2}, m_{b2}, m_{b3}, m_{b3}, I_m, I_1, I_2, I_{12}, I_{21}, I_{22}) \tag{2}$$

where m_{bi} is the mass of the block i following the direction x, y.

$K(t)$ includes the bearings stiffness k_{xi}, k_{yi} (i = 1,…,3), the shafts torsional stiffness K_i (i = 1,…,3) and the time varying gearmesh stiffness $k(t)$. It is expressed by

$$[K(t)] = \begin{bmatrix} k_{x1} & 0 & 0 & 0 & 0 & 0 & 0 & 0 & 0 & 0 & 0 & 0 \\ 0 & k_{y1} & 0 & 0 & 0 & 0 & 0 & 0 & 0 & 0 & 0 & 0 \\ 0 & 0 & s_3k(t)+k_{x2} & s_5k(t) & -s_3k(t) & -s_5k(t) & 0 & 0 & 0 & s_7k(t) & s_9k(t) & 0 \\ 0 & 0 & s_5k(t) & s_4k(t)+k_{y2} & -s_5k(t) & -s_4k(t) & 0 & 0 & 0 & s_6k(t) & s_8k(t) & 0 \\ 0 & 0 & -s_3k(t) & -s_5k(t) & s_3k(t)+k_{x3} & s_5k(t) & 0 & 0 & 0 & -s_7k(t) & -s_9k(t) & 0 \\ 0 & 0 & -s_5k(t) & -s_4k(t) & s_5k(t) & s_4k(t)+k_{y3} & 0 & 0 & 0 & -s_6k(t) & s_8k(t) & 0 \\ 0 & 0 & 0 & 0 & 0 & 0 & K_1 & -K_1 & 0 & 0 & 0 & 0 \\ 0 & 0 & 0 & 0 & 0 & 0 & -K_1 & k_f+K_1 & -k_f & 0 & 0 & 0 \\ 0 & 0 & 0 & 0 & 0 & 0 & 0 & -k_f & k_f+K_2 & -K_2 & 0 & 0 \\ 0 & 0 & s_7k(t) & s_6k(t) & -s_7k(t) & -s_6k(t) & 0 & 0 & -K_2 & s_{10}k(t)+K_2 & s_{12}k(t) & 0 \\ 0 & 0 & s_9k(t) & s_8k(t) & -s_9k(t) & s_8k(t) & 0 & 0 & 0 & s_{12}k(t) & s_{11}k(t)+K_3 & -K_3 \\ 0 & 0 & 0 & 0 & 0 & 0 & 0 & 0 & 0 & 0 & -K_3 & K_3 \end{bmatrix} \tag{3}$$

The constants s_i (i = 1,…,12) are given in the Table 1.

Table 1. Coefficients s_i of $K(t)$.

$s_1 = sin(\alpha)$	$s_7 = r_{12}.sin(\alpha)$
$s_2 = cos(\alpha)$	$s_8 = r_{21}.cos(\alpha)$
$s_3 = sin(\alpha)$	$s_9 = r_{21}.sin(\alpha)$
$s_4 = cos(\alpha)$	$s_{10} = r_{12}^2$
$s_5 = sin(\alpha).cos(\alpha)$	$s_{11} = r_{21}^2$
$s_6 = r_{12}.cos(\alpha)$	$s_{12} = r_{12}.r_{21}$

The generalized coordinate's vector of the linear dynamic model includes 12 degrees of freedom and can be defined by:

$$\{q\} = \{x_1 \quad y_1 \quad x_2 \quad y_2 \quad x_3 \quad y_3 \quad \theta_m \quad \theta_1 \quad \theta_2 \quad \theta_{12} \quad \theta_{21} \quad \theta_{22}\} \tag{4}$$

2.3 Modeling of Eccentricity Defect on the Gear System

Two typical types of mounting errors of a gear pair are eccentricities and misalignments. The eccentricity of a gear-12 is defined as the distance e_{12} between the center of rotation and the center of inertia.

In the presence of the eccentricity defect, an additional terms $U_{x2}(t)$ and $U_{y2}(t)$ are added in the expression of the displacement of the second bearing along x and y directions and can be expressed by:

$$\begin{cases} U_{x2} = e_{12}\cos(\Omega_{12}t - \lambda_{12}) \\ U_{y2} = e_{12}\sin(\Omega_{12}t - \lambda_{12}) \end{cases} \tag{5}$$

where e_{12}, Ω_{12}, λ_{12} are the eccentricity defect value, the speed of the gear-12 and the initial phase of the eccentricity defect respectively (Fig. 2).

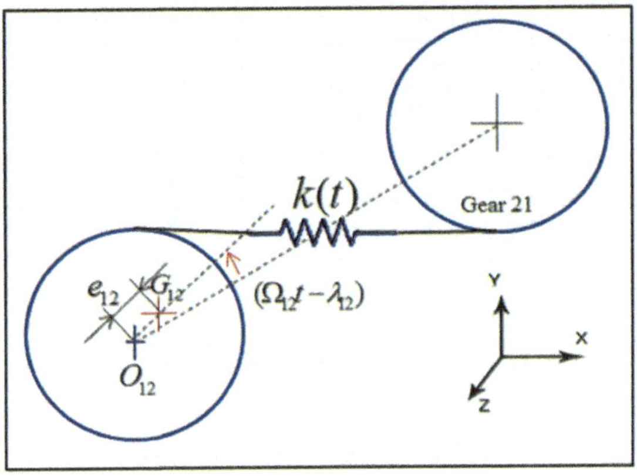

Fig. 2. Eccentricity defect model on the gear system

The total displacement of the first block can be written as follows:

$$\overrightarrow{d_{tot}} = (x_2 + U_{x2})\overrightarrow{X} + (y_2 + U_{y2})\overrightarrow{Y} \tag{6}$$

3 Results and Discussion

The time-invariant system is considered to investigate the modal properties. The meshing stiffness is considered to be constant and equal to their mean values. The i-th eigen solution associated with Eq. (7) can be obtained by solving the equation

$$\omega_i^2 M\phi_i = \tilde{K}.\phi_i \tag{7}$$

where ω_i and ϕ_i are the i-th natural frequency and the corresponding vibration mode respectively.

Based on the parameters given in Table 2, the eigen solution properties are studied by the present model. The vibration modes of the clutch-geared system can be classified into two categories including rotational and axial mode, translational mode. The natural frequencies associated with each mode are computed and listed in Table 3.

Table 2. Values of parameters for numerical studies of the drivetrain system.

Parameters	Values
Masses (kg)	m_{b1} = 4.17, m_{b2} = 2, m_{b3} = 3
Inertias (kg m^2)	I_1 = 0,02, I_2 = 12.10^{-3}, I_3 = 8.10^{-4}, I_4 = 2.10^{-4}
Torsional stiffness (N m/rad)	K_1 = K_2 = K_3 = 3.10^5, k_f = 8.10^4
Bearing stiffness (N/m)	k_{xi} = k_{yi} = k_{zi} = 10^8
Number of teeth	Z_{12} = 60, Z_{21} = 84

Table 3. Natural frequencies and vibration mode.

Vibration mode type	Natural frequencies (Hz)
Rotational mode	f_3= 77, f_4= 77, f_5= 419, f_9= 1432, f_{10}= 2563, f_{11}= 4477
Translational mode	f_1= 0, f_2= 51, f_6= 525, f_7= 987, f_8= 1223, f_{12}= 8497

In order to determine the deformations and terms of dominant movement of each body at critical speeds (which excite the natural frequencies), one has to calculate the modal kinetic energy and the modal strain energy distributions. The total modal strain energy can be expressed as follows (Hammami et al. 2015):

$$Ep_\phi = \frac{1}{2}\phi_i^t \tilde{K} \phi_i = \sum Ep_{\phi i} + Ep_{\phi k_1} \tag{8}$$

where $Ep_{\text{ø}i}$ are the strain energies of the torsional and axial stiffness. $Ep_{\text{ø}k1}$ is the strain energy of stage meshing.

The modal kinetic energy can also be defined as:

$$Ec_\phi = \frac{1}{2}\omega_i^2 \phi_i^t M \phi_i = \sum Ec_{\phi i} \tag{9}$$

where $Ec_{\text{ø}i}$ are the strain energy for each DOF in the rotational and translational movements (i = 1..$nDOF$).

The modal kinetic energy for the second bearing in the case without defect and with defect is written by Eqs. (10) and (11).

$$Ec_h = \frac{1}{2} \cdot m_{b2} \cdot \left(\dot{x}_2^2 + \dot{y}_2^2\right) \tag{10}$$

$$Ec_{ecc} = \frac{1}{2} \cdot m_{b2} \cdot \left(\dot{x}_2^2 + \dot{U}_{x2}^2 + \dot{y}_2^2 + \dot{U}_{y2}^2 + 2\dot{x}_2^2\, \dot{U}_{x2}^2 + 2\dot{y}_2^2\, \dot{U}_{y2}^2 \right) \tag{11}$$

Figure 3 shows the distribution of the modal kinetic energies for the 12 natural frequencies. The height of each column represents the percentage value of modal energy. The contribution of each degree of freedom is presented on the X-axis.

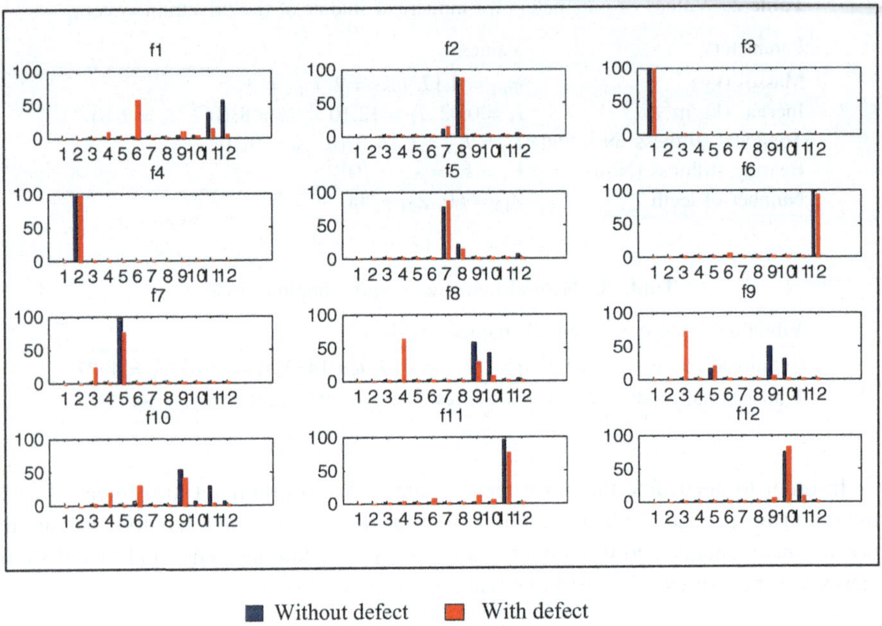

Fig. 3. Modal kinetic energies

In case without eccentricity defect, the dominant kinetic energy in the translational mode f_8 is the rotation of the clutch plate, and for the case with a defect, the dominant energy is the translation of the second bearing along y direction. For other modes such as the modes f_2, f_3 and f_{11}, the kinetic energy correspond the same degree of freedom in the two cases is the dominant but the distribution percentage is different.

The distribution of modal strain energies is shown in Fig. 4 in each frequency mode. In the case without defect, the dominant strain energy in the translational mode f_2 = 51 Hz is located in the second bearing and with the presence of defect the first bearing has the dominant strain energy. For other frequency mode, the location of the dominant energy remains the same, but with a variation of value percentage.

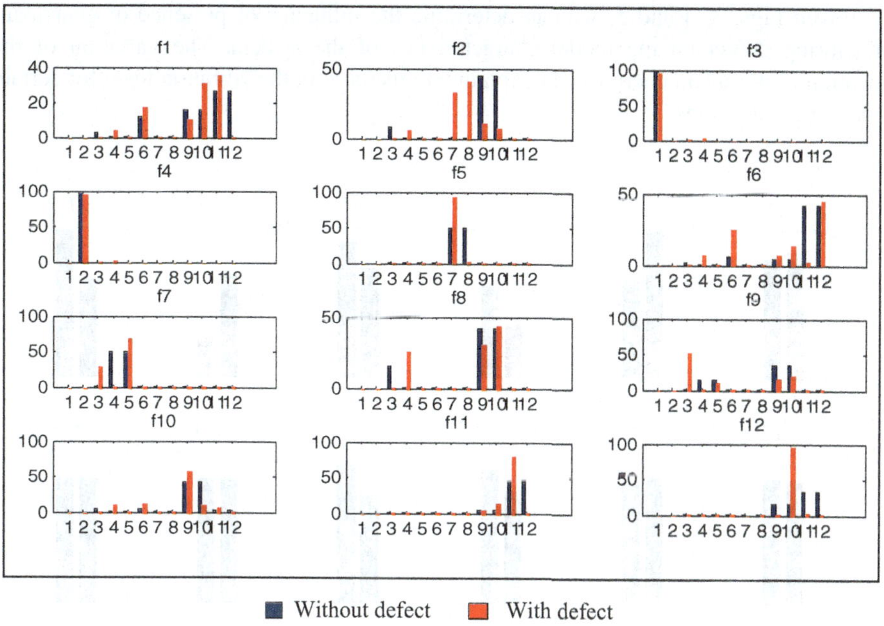

Fig. 4. Modal strain energies

In the following, we are interested in studying the influence of eccentricity defect at the level of the gear-12 on the potential energy of engagement. The total strain energy of the engagement contact will be defined by:

$$Ep_h = \frac{1}{2}.k(t).\delta^2(t) \tag{12}$$

Taking into account the defect, this energy is written in the following form:

$$Ep_{ecc} = \frac{1}{2}.k(t).\left(\delta^2(t) + e_{12}^2(t) + 2.\delta(t).e_{12}(t)\right) \tag{13}$$

where δ(t) is the transmission error and is defined by:

$$\delta(t) = (x_2 - x_3)\sin\alpha + (y_2 - y_3)\cos\alpha + \theta_{12}.r_{12} + \theta_{21}.r_{21} \tag{14}$$

The modal strain energies percentage for the engagement contact of the single stage spur gear is shown in Fig. 5. In the X-axis the case without defect is represented by 1 and the defected case is represented by 2.

The variation of the energy values due to the presence of a fault is important, and the effect of the dynamic transmission error sign is significant in the comparison of the two cases.

From Figs. 3, 4 and 5, we can determine the influence of presence of gear manufacturing defect on the modal characteristics of the system. The variation of the location of dominant energies can explain any increase in the vibration level for certain operating conditions.

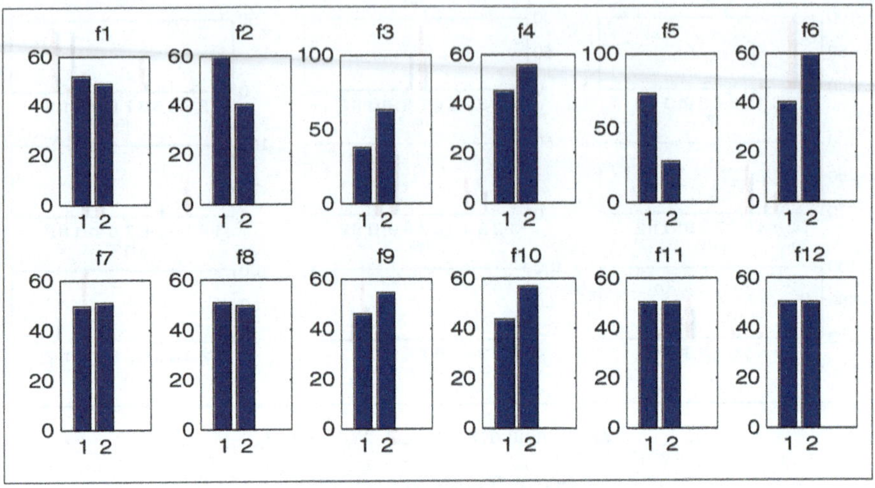

(1: without defect, 2: with defect)

Fig. 5. Modal strain energies for the engagement contact

4 Conclusion

In this paper, an analytical model for combined clutch-spur gear system was developed. An eccentricity defect on gear transmission system was included to investigate their influence on the system. Solving the eigenvalue problem allowed recovering the modal characteristics of the system. The vibration modes can be classified into rotational and translational mode. The calculation of modal kinetic and strain energies for each vibration mode will play an important role in defect sensitivity analyses.

References

Fakhfakh, T., Walha, L., Louati, J., Haddar, M.: Effect of manufacturing and assembly defects on two-stage gear systems vibration. Int. J. Adv. Manufact. Technol. **29**, 1008–1018 (2006)

Walha, L., Driss, Y., Fakhfakh, T., Haddar, M.: Effect of manufacturing defects on the dynamic behaviour for a helical two-stage gear system. Mécanique & Industrie **10**, 365–376 (2009)

Gaillard, C.L., Singh, R.: Dynamic analysis of automotive clutch dampers. Appl. Acoust. **60**, 399–424 (2000)

Duan, C., Singh, R.: Dynamics of a 3dof torsional system with a dry friction controlled path. J. Sound Vib. **289**, 657–688 (2006)

Walha, L., Driss, Y., Khabou, M.T., Fakhfakh, T., Haddar, M.: Effects of eccentricity defect on the nonlinear dynamic behavior of the mechanism clutch-helical two stage gear. Mech. Mach. Theory **46**, 986–997 (2011)

Caruntu, C.F., Lazar, M., Di Cairano, S.: Driveline oscillations damping: a tractable predictive control solution based on a piecewise affine model. Nonlinear Anal. Hybrid Syst. **19**, 168–185 (2016)

Ghorbel, A., Abdennadher, M., Zghal, B., Walha, L., Haddar, M.: Modal analysis and dynamic behavior for analytical drivetrain model. J. Mech., 1–17 (2017)

Ghorbel, A., Abdennadher, M., Walha, L., Zghal, B., Haddar, M.: Vibration Analysis of a Nonlinear Drivetrain System in the Presence of Acyclism. In: International Conference Design and Modeling of Mechanical Systems, pp. 541–550. Springer, Cham, March 2017

Hammami, A., Del Rincon, A.F., Rueda, F.V., Chaari, F., Haddar, M.: Modal analysis of back-to-back planetary gear: experiments and correlation against lumped-parameter model. J. Theor. Appl. Mech. **53**(1), 125–128 (2015)

Estimation of Road Disturbance for a Non Linear Half Car Model Using the Independent Component Analysis

Dorra Ben Hassen[1(✉)], Mariem Miladi[1], Mohamed Slim Abbes[1], S. Caglar Baslamisli[2], Fakher Chaari[1], and Mohamed Haddar[1]

[1] Mechanics, Modeling and Production Laboratory, National Engineering School of Sfax (ENIS), BP 1173, 3038 Sfax, Tunisia
dorra.benhassen@yahoo.fr, mariam.mi@hotmail.fr, ms.abbes@gmail.com, fakher.chaari@gmail.com, Mohamed.haddar@enis.rnu.tn

[2] Department of Mechanical Engineering, Hacettepe University, Beytepe, 06800 Ankara, Turkey
caglar.baslamisli@gmail.com

Abstract. The identification of the road profile disturbance acting on a vehicle was the objective of many recent researches. This estimation remains very interesting since it contributes to study the dynamic behavior of the vehicle in one side and to choose a control law later in other side. However most of the used techniques have many drawbacks such us those based on direct measurements of the profile which need costly profilometers or those based on neural network algorithm which are very complicated. So the purpose of this research is to use a new method named the Independent Component Analysis (ICA) to estimate the road profile. This method is based on the so-called inverse problem. So it necessitates only the knowledge of the dynamic responses of the vehicle to identify the road disturbance. Therefore the Newmark algorithm is used in this paper to extract the dynamic responses of the system under study which is a non linear half car model. Starting from these responses, the ICA algorithm is applied. The validation of the obtained results is done using some performance criteria which are the relative error and the MAC number. Finally a good agreement is found between the original profile and the estimated one.

Keywords: Non linear half car model · Road disturbance · ICA

1 Introduction

Vehicle dynamics is a domain of considerable interest for many years. It encompasses the intervention of many factors: driver, vehicle and loads (Rill 2004). Many models of vehicle are used in order to determine their dynamic behaviour. Such as E. Duni (Duni et al. 2003), in his studies uses a finite element method in order to simulate the dynamic response of a full vehicle model subjected to different types of road excitations. Others implement a bicycle model with four degree of freedom (Hunt 1989; Mavros 2008) and they concluded that the characteristics of the road profile influence on the dynamic

A. Felkaoui et al. (Eds.): SIGPROMD'2017, ACM 12, pp. 96–103, 2019.
https://doi.org/10.1007/978-3-319-96181-1_9

response of the system. Pacejka (2005) also focuses on the study of the handling behaviour of a bicycle model with transient tyres. In this paper a non linear half car model is studied in order to identify the road disturbance. This identification was done using different techniques such as direct measurements (Kim et al. 2002), but its cost is very expensive. Other researchers use the estimation algorithms (Solhmirzaei et al. 2012), however they necessitate a long computing time (Fauriat et al. 2016). So in this study the proposed method of road profile estimation is the ICA. It is used to estimate the excitation force in many studies (Dhief et al. 2016; Hassen et al. 2017; Taktak et al. 2012). This method is simple to apply and permits to identify the road excitation in real time.

This paper is structured as follows: the first part present the studied system and its mathematical formulation. Then the results obtained by the ICA are presented in the second part and finally a good agreement between the original excitation and the estimated one is obtained.

2 Half Car Model

The dynamic model (Meywerk 2015) of the half car is presented in Fig. 1.

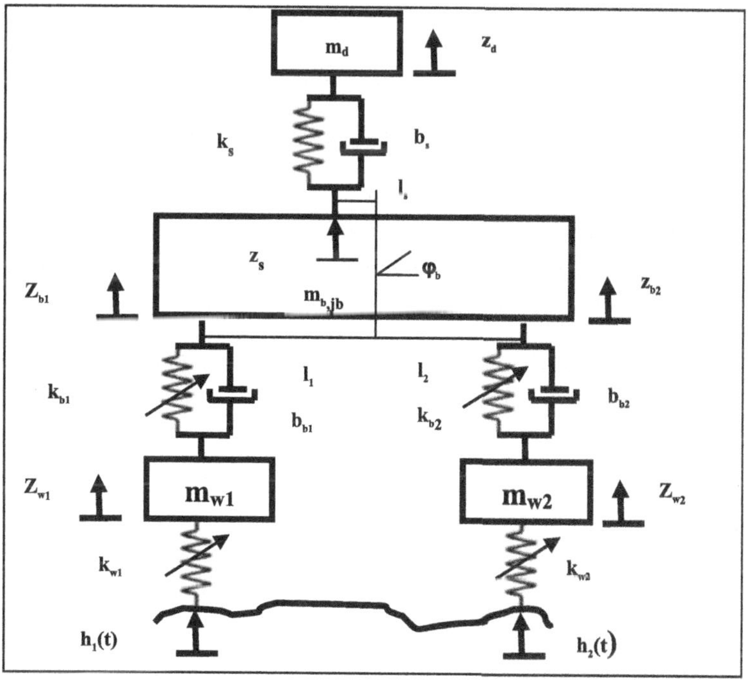

Fig. 1. Full vehicle model

This model has five degree of freedom:

m_{w1} and m_{w2} are the masses of the wheels. They are attached to the road via two non linear springs k_{w1} and k_{w2}. Their deflections are noted z_{w1} and z_{w2}.

The vertical displacement of the two suspension systems are noted respectively z_{b1} and z_{b2}. These suspension have a non linear behaviours due to the non linear stiffness k_{b1} and k_{b2}. The dampers are noted b_{b1} and b_{b2}.

Z_b and φ_b denote respectively the displacement of the center of gravity and the pitch angle.

The vertical displacement of the human's seat is noted z_d.

zb_1 and zb_2 can be expressed in function of the variable z_d as (Meywerk 2015):

$$z_{b1} = z_b - l_1\varphi_b \tag{1}$$

$$z_{b2} = z_b + l_2\varphi_b \tag{2}$$

And the coordinate z_s is expressed in terms of z_b as follow:

$$z_s = z_b - l_s\varphi_b \tag{3}$$

For the non linearity it's expressed by the following expressions:

$$F_{b1} = k_{b1}\,\Delta l + \beta_1 k_{b1}\,\Delta l^2 + \beta_2 k_{b1}\Delta l^3 \tag{4}$$

and

$$F_{b2} = k_{b2}\,\Delta l + \beta_1 k_{b2}\,\Delta l^2 + \beta_2 k_{b2}\Delta l^3 \tag{5}$$

With:

Δl is the difference between the two displacements z_{b1} and z_{w1} in Eq. (4)

And Δl is the difference between the two displacements z_{b2} and z_{w2} in Eq. (5).

β_1, β_2 are two non linear constants.

$\beta_1 = 0.1$ *and* $\beta_2 = 0.4$

The tire is modeled as a spring with a non linear stiffness k_2 in parallel with a linear damper c_2. The expression of the non linear tire stiffness is taken from Li et al. (2011) as:

$$F_{w1} = k_{w1}\,\Delta l + \beta_3 k_{w1}\Delta l^2 \tag{6}$$

And

$$F_{w2} = k_{w2}\,\Delta l + \beta_3 k_{w2}\Delta l^2 \tag{7}$$

With:

Δl is the difference between the displacement kw1 and the road excitation $h_1(t)$ in Eq. (6) and Δl is the difference between the displacement kw2 and the road excitation $h_1(t)$ in Eq. (7).

β_3 is the non linear tire coefficient. Its value is taken from (Li et al. 2011):

$$\beta_3 = 0.01 \tag{8}$$

To solve this non linear system, the implicit schema of Newmark coupled with Newton Raphson Method was used using the parameters presented in the following Table 1:

Table 1. Parameters of the full vehicle model

Parameters	Variable value	Variable unit
Mass of the chassis	$m_b = 960$	[Kg]
Mass of the tires	$m_{w1} = m_{w2} = 36$	[Kg]
Suspension stiffness	$k_{b1} = k_{b2} = 16000$	[N/m]
Tire stiffness	$K_{w1} = k_{w2} = 10^5$	[N/m]
Suspension damping	$bb_1 = bb_2 = 100$	[N/ms]
Driver's mass	$m_d = 90$	[Kg]
Moment of inertia	$Jb = 500$	[Kg/m^2]
Driver seat's rigidity	$k_s = 2000$	[N/m]
Driver seat's damping	$b_s = 10$	[N/ms]
l_1	$l_1 = 1.8$	[m]
l_2	$l_2 = 0.8$	[m]

Concerning the road excitation, we take in the first wheel a bump excitation and in the second the same excitation with a short delay as presented below:

$$h_1(t) = \begin{Bmatrix} 0.05\,\frac{1-\cos(8\pi t)}{2} & if\,1 \le t \le 1.25 \\ 0.05\,\frac{1-\cos(8\pi t)}{2} & if\,5 \le t \le 5.25 \\ 0\ otherwise & \end{Bmatrix}$$

$$h_2(t) = \begin{Bmatrix} 0.05\,\frac{1-\cos(8\pi t)}{2} & if\,1.25 \le t \le 1.5 \\ 0.05\,\frac{1-\cos(8\pi t)}{2} & if\,5.25 \le t \le 5.5 \\ 0\ otherwise & \end{Bmatrix}$$

The following figure presents the two excitations applied on the wheels (Fig. 2):

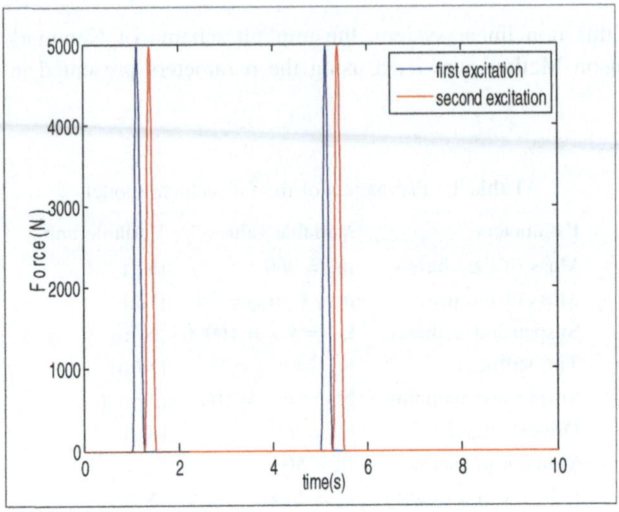

Fig. 2. Bump excitations

3 Description of the Applied Algorithm: ICA

The ICA is a method which aims to decompose a random signal X in independent components statistically (Abbes et al. 2011; Dhief et al. 2016).

The vector X can be written as (Hassen et al. 2017)

$$X(t) = [A]\{S\} \tag{9}$$

where:

A: Mixing matrix
S: Vector of source signals.

The task of ICA is to estimate A and S based only on the knowledge of the vector X. This estimation requires some assumptions:

- The components of the vector S must be statistically independent
- The number of the observed signals is equal to the number of the estimated sources.
- The components of the vector S must have a non-Gaussian distribution.

By validating these assumptions, the ICA define each column of the matrix A and after that compute the separating matrix W such as:

$$W = A^{-1} \tag{10}$$

Then the ICA estimate the corresponding source signal defined by:

$$\{S\} = [W]\{X\} \tag{11}$$

Finally, the vector X undergoes some pretreatments (it must be centered and whitened) to have a successful separation.

4 Numerical Results

Starting from the observed signals presented by Fig. 3, the ICA is applied to the half car model in order to reconstruct the original excitations. We added a Gaussian random noise with zero mean value and a standard deviation σ equal to 0.5 (Akrout et al. 2012) on the observed signals in order to study the efficiency of the ICA.

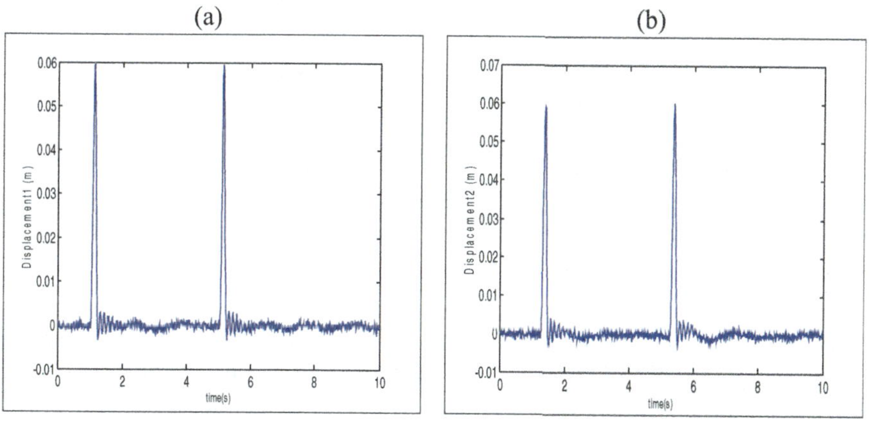

Fig. 3. Observed signals (a) displacement of X1 (b) displacement of X2

The results of the ICA are presented by the following figures (Fig. 4).

We note that the ICA can identify the original signals. There is a small delay and perturbation due to the effect of the non linearity and the noise added to the sensors. But the obtained results remain in agreement with the original ones. The following table resumes the performance criteria (Table 2).

We can note that Mac value is near to one for the two studied signals, also the error has minimum value. These results confirm that the ICA is able to identify the original signal.

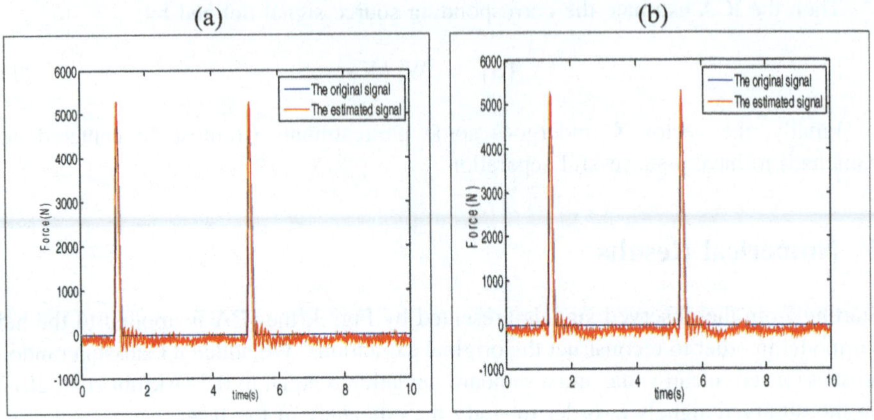

Fig. 4. Identification of the road profile by the ICA (a) excitation 1 (b) excitation2

Table 2. Validation of the results

	Mac	Relative error (%)
Profile 1	0.93	2.5071
Profile 2	0.92	2.5

5 Conclusion

This paper deals with the application of the ICA in order to reconstruct the road excitations. This method is applied to a non linear half car model. And the obtained results are in concordance with the original sources even with the non linear case.

This will be of a good importance to study the dynamic behavior of the system and to choose the adequate controller in future work.

References

Rill, G.: Vehicle Dynamics. University of applied sciences, Regensburg (October 2004)

Duni, E., Monfrino, G., Saponaro, R., Caudano, M., Urbinati, F., Marco, S., Antonino, P.: Numerical simulation of full vehicle dynamic behaviour based on the interaction between ABAQUS/Standard and explicit codes. In: Abaqus Users' Conference, June, Munich (2003)

Hunt, H.E.M.: Stochastic modelling of vehicles for calculation of ground vibration. In: Proceedings of the 11th IAVSD-Symposium-The Dynamics of Vehicles on Roads and Tracks, vol. 18. Taylor & Francis, London, August 1989

Mavros, G.: A study on the influences of tyre lags and suspension damping on the instantaneous response of a vehicle. Proc. Inst. Mech. Eng. Part D J. Automob. Eng. **222**(4), 485–498 (2008)

Pacejka, H.: Tire and Vehicle Dynamics. Elsevier, Amsterdam (2005)

Kim, H.J., Yang, H.S., Park, Y.P.: Improving the vehicle performance with active suspension using road-sensing algorithm. Comput. Struct. **80**(18), 1569–1577 (2002)

Solhmirzaei, A., Azadi, S., Kazemi, R.: Road profile estimation using wavelet neural network and 7-DOF vehicle dynamic systems. J. Mech. Sci. Technol. **26**(10), 3029–3036 (2012)

Fauriat, W., Mattrand, C., Gayton, N., Beakou, A., Cembrzynski, T.: Estimation of road profile variability from measured vehicle responses. Veh. Syst. Dyn. **54**(5), 585–605 (2016)

Dhief, R., Taktak, M., Tounsi, D., Akrout, A., Haddar, M.: Application of the independent components analysis in the reconstruction of acoustic sources in duct systems. Arab. J. Sci. Eng., 1–10 (2016)

Hassen, D.B., Miladi, M., Abbes, M.S., Baslamisli, S.C., Chaari, F., Haddar, M.: Application of the operational modal analysis using the independent component analysis for a quarter car vehicle model. In: Advances in Acoustics and Vibration (pp. 125–133). Springer, Berlin (2017)

Taktak, M., Tounsi, D., Akrout, A., Abbès, M.S., Haddar, M.: One stage spur gear transmission crankcase diagnosis using the independent components method. Int. J. Veh. Noise Vib. **8**(4), 387–400 (2012)

Meywerk, M.: Vehicle Dynamics. Wiley, London (2015)

Li, S., Lu, Y., Li, H.: Effects of parameters on dynamics of a nonlinear vehicle-road coupled system. JCP **6**(12), 2656–2661 (2011)

Abbès, M.S., Akrout, M.A., Fakhfekh, T., Haddar, M.: Vibratory behavior of double panel system by the operational modal analysis. Int. J. Model. Simul. Sci. Comput. **2**(4), 459–479 (2011)

Akrout, A., Tounsi, D., Taktak, M., Abbès, M.S., Haddar, M.: Estimation of dynamic systems excitation forces by the independent component analysis. Int. J. Appl. Mech. **4**(3), 1250032 (2012)

Transfer Path Analysis of Planetary Gear with Mechanical Power Recirculation

Ahmed Hammami[1,2(✉)], Alfonso Fernandez del Rincon[2], Fakher Chaari[1], Fernando Viadero Rueda[2], and Mohamed Haddar[1]

[1] Laboratory of Mechanics, Modeling and Production (LA2MP), National School of Engineers of Sfax, BP 1173, 3038 Sfax, Tunisia
ahmed.hammami2109@gmail.com, fakher.chaari@gmail.com, mohamed.haddar@enis.rnu.tn

[2] Department of Structural and Mechanical Engineering, Faculty of Industrial and Telecommunications Engineering, University of Cantabria, Avda de los Castros s/n, 39005 Santander, Spain
{alfonso.fernandez, fernando.viadero}@unican.es

Abstract. Planetary gears can transmit higher power density levels because they use multiple power paths formed by each planet branches. In order to study the propagation of vibration between components of planetary gear test bench with mechanical power recirculation, an approach to the classical transfer path analysis (TPA) method is used to improve vibration control of planetary gear test bench. This approach termed Global Transmissibility Direct Transmissibility (GTDT) avoids the drawbacks of the classical TPA which are decoupling of the active part in the measurements of the Frequency Response Functions (FRFs) of source-receiver paths and the difficult measurement of the operational forces. The Global Transmissibility Direct Transmissibility (GTDT) is two steps method: the first step is the measurements of transmissibility which requires no disassembly tests and the second step is the measurement of the operational responses which is easier than measurement of the operational forces. In fact, tri-axial accelerometers are mounted in each component of the back-to-back planetary gear test bench and the transfer functions (frequency response functions, FRFs) are measured using hammer impact test in order to measure the global transmissibilities. Then, the direct transmissibilities are computed from the global transmissibilities. Finally, reconstructed operation responses are shown in the partial path contribution (PPC) plots to compare the vibration level of each component and to know its contribution in the transfer of vibration.

Keywords: Planetary gear · Transfer path analysis
Global Transmissibility Direct Transmissibility · Mechanical power recirculation

1 Introduction

The transfer path analysis (TPA) is required to improve vibration and noise control. It was firstly is used in the automotive industry in order to analyze the different contributions of vibration and noise applied to the driver and passenger positions (Plunt 2005). This celebrated technique can be applied into two steps which are the

A. Felkaoui et al. (Eds.): SIGPROMD'2017, ACM 12, pp. 104–115, 2019.
https://doi.org/10.1007/978-3-319-96181-1_10

measurements of the Frequency Response Functions (FRFs) of source-receiver paths using an artificial excitation and the measurement of the operational forces or volume velocities acting on the system.

In the classical TPA, the global system is divided into an active part containing the sources and a passive part containing the transfer path and receiver points.

This method has two drawbacks which are the difficulties of decoupling the active part of the system in the measurements of FRFs and the difficulties to measure the operational forces.

In order to avoid the limits of the classical TPA, many approaches are proposed: a one step method, called Operational Transfer Path Analysis OTPA, was introduced by Rebeiro et al. (2000). This method uses a matrix of transmissibility and it requires measurement data of the operating vehicle in order to perform the analysis. The OPTA allows the identification of the transfer paths without disassembling the system and facilitates the measure of the operational forces (De Sitter et al. 2010) (De Klerk and Ossipov 2010). This approach has several technical drawbacks: transmissibility is different from transfer functions and the transmissibility matrix concept should carefully be used because the input measurement points are inherently coupled (Gajdatsy et al. 2010).

The OPTA method was improved by Roozen and Leclère (2013) by using a non instrumented excitation on small gearbox in two steps: the first step is to identify the transmissibility matrix: the system is excited by hammer strokes which are independent. The second step is to measure the operational forces through force transducers integrated in the pins which supported the gearbox. Janssens et al. (2011) proposed another approach so-called OPAX. It combines the measurements of the transfer path with operational measurements. In this method, force sensors are mounted on the connection between the source and the receiver. The drawback of the last two approaches is the difficulties of the measurement of the operational forces in many cases.

In order to avoid these limits, another two steps strategy was developed by Magrans (1981): the first step is the measurements of transmissibility which requires no disassembly tests and the second step is the measurement of the operational responses. In contrast, the so-called direct transmissibilities obtained from the measured transmissibilities are used for the operational response contribution. The direct transmissibilities approach is used also by Guash and Magrans (2004) (Guasch 2009) and termed Global Transmissibility Direct Transmissibility (GTDT).

This paper will be dedicated to the last transfer path analysis approach termed Global Transmissibility Direct Transmissibility (GTDT) where computation of the direct transmissibility from the global transmissibility is explained and the reconstruction of any degree-of-freedom is described. Then, the experimental setup and measurements are detailed. Experimental results are shown for system running under stationary conditions.

2 Description of the Test Bench

The test bench is composed of two identical planetary gear sets (Fig. 1): The first planetary gear is a “test gear set” and the second planetary gear is a “reaction gear set” which has the same gear ratio and inject the output power to the input allowing the

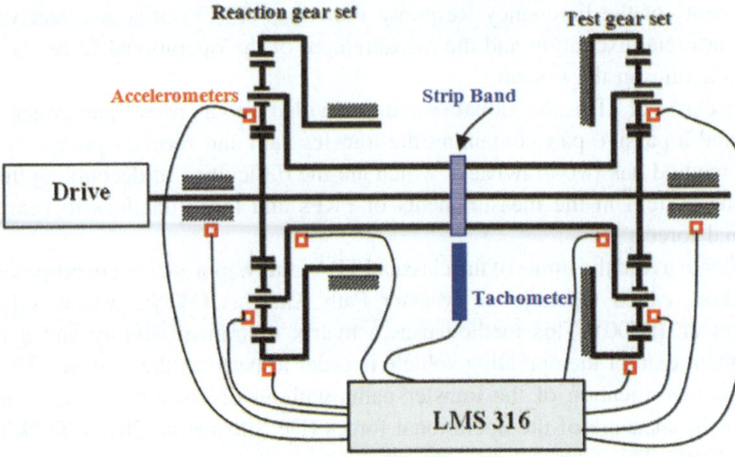

Fig. 1. Instrumentation layout

mechanical power circulation (Hammami et al. 2015a, b). The two planetary gears are connected back-to-back: the sun gears of both planetary gear sets are connected through a common shaft and the carriers of both planetary gear sets are connected to each other through a rigid hollow shaft (Hammami et al. 2016).

Fig. 2. Accelerometers mounted in components of the planetary gear test bench

An optic tachometer (Compact VLS7) which is placed along the hollow carriers' shaft measure its instantaneous angular velocity.

In order to measure the global transmissibilities, tri-axial accelerometers are mounted in each component of the back-to-back planetary gear (Fig. 2).

The wires from accelerometers are connected to the acquisition system "LMS SCADAS 316 system".

The data will be processed with the software "LMS Test.Lab" and the transfer functions (frequency response functions, FRFs) are recorded using hammer impact test.

3 Numerical Results

The Global Transmissibilities Direct Transmissibilities approach is used to study the propagation of vibration and sound. This method allows avoiding the drawbacks of the classical transfer path analysis and it is applied in two steps: the first step is the measurements of transmissibility which requires no demounting tests and the second step is the measurements of the operational responses. In contrast, the so-called direct transmissibilities obtained from the measured transmissibilities are used to compute the operational response contribution.

3.1 Calculation of the Direct Transmissibility Matrix

Using the classical TPA, the operational response at any degree-of-freedom x_i is a function of the external load acting on the system f_j, the number of the external loads acting the system N_f and the transfer function (FRFs) H_{ij}:

$$x_i = \sum_{j=1}^{N_f} H_{ij} f_j \tag{1}$$

For the GTDT reconstruction of the response of x_i:

$$x_i = \sum_{j=1, j \neq i}^{N-1} T_{ij}^D x_j + T_{ii}^D x_i^{ext} \tag{2}$$

The response at any degree-of-freedom x_i is decomposed into contributions from other degree-of-freedom responses and the portion of that the external load is directly acting on it ($x_i^{ext} = 0$ if no force on it). T_{ij}^D and T_{ii}^D are the direct or blocked transmissibilities.

The blocked transmissibility is computed from the global transmissibility T_{ij}^G which corresponds to the ratio between the response of the ith degree-of-freedom and the response of the jth degree-of-freedom when j is excited:

$$T_{ij}^G = \frac{x_j}{x_i} = \frac{H_{ij} f}{H_{ii} f} = \frac{H_{ij}}{H_{ii}} \tag{3}$$

The matrix form of the global transmissibilities is:

$$T^G = diag(\frac{1}{H_{11}}, \ldots, \frac{1}{H_{nn}}) \times H \tag{4}$$

The inverse matrix of T^G is written:

$$T^{G^{-1}} = H^{-1} \times diag(H_{11}, \ldots, H_{nn}) = Z \times diag(H_{11}, \ldots, H_{nn}) \tag{5}$$

The direct transmissibility corresponds to the ratio between the response of the ith degree-of-freedom and the response of the jth degree-of-freedom when j is excited and all remaining degree-of-freedom, except i and j, are blocked. In this case and as the system is governed by $Zx = f$ (Guasch et al. 2013):

$$\begin{bmatrix} Z_{ii} & Z_{ij} \\ Z_{ji} & Z_{jj} \end{bmatrix} \begin{Bmatrix} x_i \\ x_j \end{Bmatrix} = \begin{Bmatrix} 0 \\ 1 \end{Bmatrix} \tag{6}$$

From the Eq. (6), the responses in i^{th} and j^{th} degree-of-freedom can respectively be written:

$$x_i = \frac{Z_{ij}}{Z_{ij}Z_{ji} - Z_{ii}Z_{jj}} \tag{7}$$

$$x_j = \frac{-Z_{ii}}{Z_{ij}Z_{ji} - Z_{ii}Z_{jj}} \tag{8}$$

So, the direct transmissibility is defined as:

$$T^D_{ij} = \frac{x_i}{x_j} = -\frac{Z_{ij}}{Z_{ii}} \tag{9}$$

For the direct transmissibility T^D_{ii}, all degrees-of-freedom are blocked except the ith d.o.f. So, $Z_{ii}x_i = 1$. Besides, only the ith d.o.f is excited. So, $H_{ii} = x_i$. Therefore:

$$T^D_{ii} = \frac{1}{Z_{ii}H_{ii}} \tag{10}$$

From Eq. (6), it can be written:

$$T^{G^{-1}}\Big|_{ij} = Z_{ij}H_{ii} \tag{11}$$

Dividing T^D_{ij} in (9) by T^D_{ii} in (10) and taking into account to (11) gives:

$$\frac{T^D_{ij}}{T^D_{ii}} = -T^{G^{-1}}\Big|_{ij}, i \neq j \tag{12}$$

Taking $i = j$ in (11) and using (12) provides:

$$\frac{1}{T_{ii}^{D}} = T^{G^{-1}}\Big|_{ii} \tag{13}$$

From (12) and (13), we have $T_{ij}^{DE} = T_{ij}^{D} = -T_{ii}^{D}T^{G^{-1}}\Big|_{ij}, \forall i \neq j$ and $T_{ii}^{DE} = -1 = -T_{ii}^{D}T^{G^{-1}}\Big|_{ii}$. Where, T^{DE} corresponds to the direct transmissibilities matrix with its diagonal replaced by -1 values. These relations can be written in the matrix form:

$$T^{DE} = -\Lambda_{TD}T^{G^{-1}} \tag{14}$$

Where $\Lambda_{TD} = diag(T_{11}^{D}, \ldots, T_{nn}^{D})$:

The computation of the direct transmissibility matrix involves the inversion of the measured global transmissibility (Eq. (14)).

3.2 Operational Response Decomposition

The operational response at any subsystem, x_i^{ext}, is due to the action of the external forces. Then, the overall operational response at any degree-of-freedom is function of the global transmissibility matrix: $x = T^{G}x^{ext}$

$$x^{ext} = T^{G^{-1}}x \tag{15}$$

As $T^{G^{-1}} = -\Lambda_{TD}^{-1}T^{DE}$, the Eq. (15) can be written:

$$-\Lambda_{TD}x^{ext} = T^{DE}x = (T^{D} - \Lambda_{TD} - I)x \tag{16}$$

The last equation is a generalisation of the Eq. (2) that includes all degree-of-freedom operational decompositions using direct transmissibilities in a single expression (Guasch and Magrans 2004) (Guasch 2009):

$$x = (T^{D} - \Lambda_{TD})x + \Lambda_{TD}x^{ext} \tag{17}$$

3.3 Experimental Setup

Twelve degrees-of-freedom corresponding to the rotational coordinates are considered for back-to back planetary gear (torsional model). x is the degree of freedom vector expressed by:

$$x = \{u_{cr}, u_{rr}, u_{sr}, u_{1r}, u_{2r}, u_{3r}, u_{ct}, u_{rt}, u_{st}, u_{1t}, u_{2t}, u_{3t}\}^{T} \tag{18}$$

The rotational coordinates are $u_{rj} = r_{rj}\theta_{rj}$ for reaction gear set $u_{tj} = r_{tj}\theta_{tj}$ and for test gear set where j = c,r,s,1,2,3. θ_{rj} and θ_{tj} are the rotational components; r_{rj} and r_{tj} are the base radius for the sun, ring and planets and the radius of the circle passing through the planets centres for the carrier.

The global transmissibilities are measured in the first step of the GTDT. The global transmissibility is defined as:

$$T^G = \begin{bmatrix} 1 & T_{cr,rr} & T_{cr,sr} & T_{cr,1r} & T_{cr,2r} & T_{cr,3r} & T_{cr,ct} & T_{cr,rt} & T_{cr,st} & T_{cr,1t} & T_{cr,2t} & T_{cr,3t} \\ T_{rr,cr} & 1 & T_{rr,sr} & T_{rr,1r} & T_{rr,2r} & T_{rr,3r} & T_{rr,ct} & T_{rr,rt} & T_{rr,st} & T_{rr,1t} & T_{rr,2t} & T_{rr,3t} \\ T_{sr,cr} & T_{sr,rr} & 1 & T_{sr,1r} & T_{sr,2r} & T_{sr,3r} & T_{sr,ct} & T_{sr,rt} & T_{sr,st} & T_{sr,1t} & T_{sr,2t} & T_{sr,3t} \\ T_{1r,cr} & T_{1r,rr} & T_{1r,sr} & 1 & T_{1r,2r} & T_{1r,3r} & T_{1r,ct} & T_{1r,rt} & T_{1r,st} & T_{1r,1t} & T_{1r,2t} & T_{1r,3t} \\ T_{2r,cr} & T_{2r,rr} & T_{2r,sr} & T_{2r,1r} & 1 & T_{2r,3r} & T_{2r,ct} & T_{2r,rt} & T_{2r,st} & T_{2r,1t} & T_{2r,2t} & T_{2r,3t} \\ T_{3r,cr} & T_{3r,rr} & T_{3r,sr} & T_{3r,1r} & T_{3r,2r} & 1 & T_{3r,ct} & T_{3r,rt} & T_{3r,st} & T_{3r,1t} & T_{3r,2t} & T_{3r,3t} \\ T_{ct,cr} & T_{ct,rr} & T_{ct,sr} & T_{ct,1r} & T_{ct,2r} & T_{ct,3r} & 1 & T_{ct,rt} & T_{ct,st} & T_{ct,1t} & T_{ct,2t} & T_{ct,3t} \\ T_{rt,cr} & T_{rt,rr} & T_{rt,sr} & T_{rt,1r} & T_{rt,2r} & T_{rt,3r} & T_{rt,ct} & 1 & T_{rt,st} & T_{rt,1t} & T_{rt,2t} & T_{rt,3t} \\ T_{st,cr} & T_{st,rr} & T_{st,sr} & T_{st,1r} & T_{st,2r} & T_{st,3r} & T_{st,ct} & T_{st,rt} & 1 & T_{st,1t} & T_{st,2t} & T_{st,3t} \\ T_{1t,cr} & T_{1t,rr} & T_{1t,sr} & T_{1t,1r} & T_{1t,2r} & T_{1t,3r} & T_{1t,ct} & T_{1t,rt} & T_{1t,st} & 1 & T_{1t,2t} & T_{1t,3t} \\ T_{2t,cr} & T_{2t,rr} & T_{2t,sr} & T_{2t,1r} & T_{2t,2r} & T_{2t,3r} & T_{2t,ct} & T_{2t,rt} & T_{2t,st} & T_{2t,1t} & 1 & T_{2t,3t} \\ T_{3t,cr} & T_{3t,rr} & T_{3t,sr} & T_{3t,1r} & T_{3t,2r} & T_{3t,3r} & T_{3t,ct} & T_{3t,rt} & T_{3t,rt} & T_{3t,1t} & T_{3t,2t} & 1 \end{bmatrix} \tag{19}$$

3.4 Experimental Results

3.4.1 Global and Direct Transmissibilities

In the first step, the measured global transmissibilities with their direct counterparts are showed. In Fig. 3, the logarithms of the squared module of the global and direct transmissibilities between test carrier and test planet 2 ($T^D_{ct,2t}$ and $T^G_{ct,2t}$) for two different angular positions are plotted. All these functions are different. As an example, $T^G_{ct,2t}$ stands for the ratio between the responses of the test planet 2 and the test carrier, when the test carrier is excited, whereas $T^D_{ct,2t}$ corresponds to the same ratio but keeping all the remaining system degree-of-freedoms fixed.

Given that test ring is not directly connected to the reaction ring, the direct (blocked)$T^D_{rr,rt}$ should be zero and hence become minus infinity in a logarithmic plot (Maia et al. 2001; Ribeiro et al. 2000). However, this is not the case for the direct transmissibility $T^D_{rr,rt}$ (Fig. 4). This behavior is explained by the fact that the test ring and the reaction ring are connected through the chassis of the test bench: The test ring is blocked to the chassis of the test bench. The reaction ring has a rotary motion around the chassis of the test bench

Also, the direct transmissibility $T^D_{rr,rt}$ should be smaller than its respective global transmissibility $T^G_{rr,rt}$.

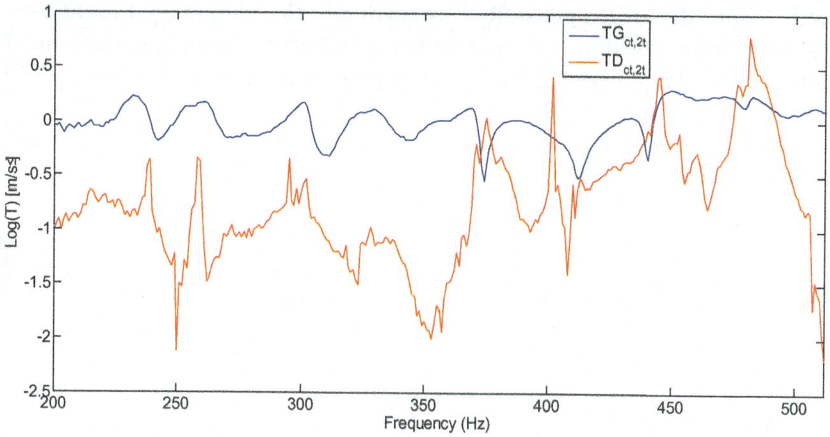

Fig. 3. Global and direct transmissibilities $T^D_{ct,2t}$ and $T^G_{ct,2t}$

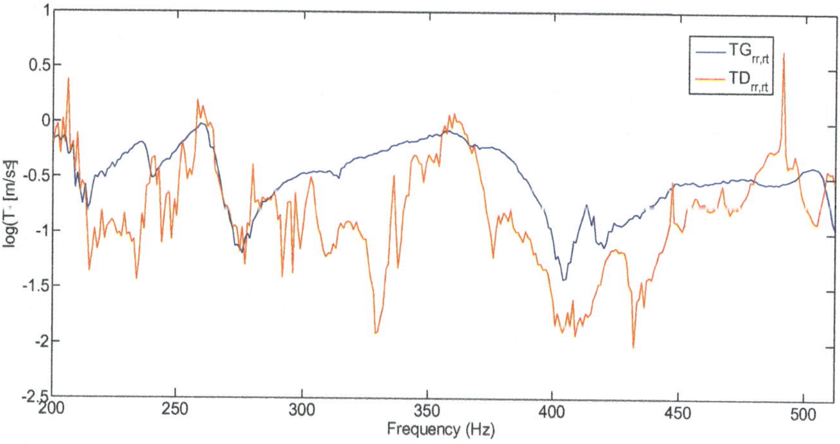

Fig. 4. Global and direct transmissibilities $T^D_{rr,rt}$ and $T^G_{rr,rt}$

The fact that the direct transmissibility is not null, when it should, can pollute the operational response reconstruction. Neglecting the error in the reconstructed operational displacement will depend on how small the involved direct transmissibility becomes with respect to all the other direct transmissibility, and on the responses of the degree-of-freedoms it connects as well.

The operational response reconstruction is studied in the stationary conditions.

3.4.2 Operational Response Reconstruction in the Stationary Conditions

The operational situation is considered under stationary conditions where the driving motor excites the system. Results correspond to partial path contribution (PPC) plots (Gajdatsy et al. 2010).

Figure 5 shows PPC surface plot of the logarithm of the squared modulus of all reconstructed operation responses.

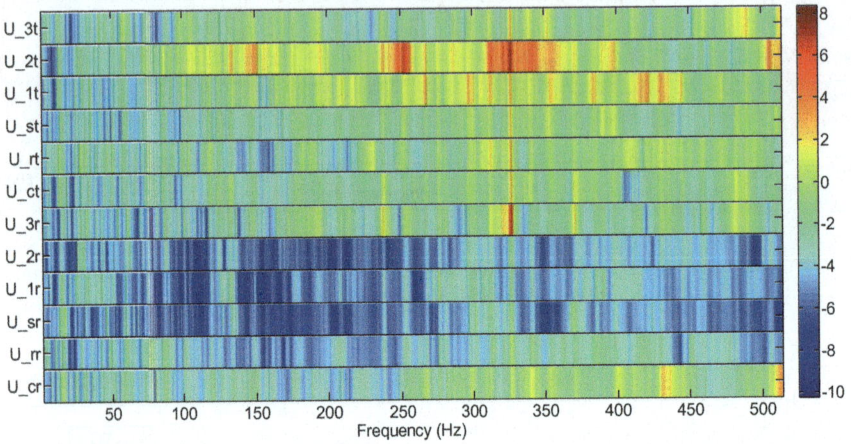

Fig. 5. PPC surface plot of the reconstructed operation responses

It is shown on the last figure that the vibration level of all reconstructed operation responses is higher around the frequency 321 Hz which correspond to the gear mesh frequency. In addition, it is clear that the second test planet presents the highest vibration level which can be explained by the fact that the test planet 2 has two pairs of teeth on contact with the sun and the ring and its transmissibility is higher. In addition, the test planet 2 has a position error on the carrier and it is preloaded before the other planets which has an effect on the load sharing behaviour (Hammami et al. 2017).

In order to understand better the higher vibration level of the second test planet, a decomposition of the reconstructed operational response is presented on the Fig. 6.

It is observed on this figure that the summation of all response contributions almost perfectly match the constructed response of the 2nd test planet. Also, all these responses are higher around the gear mesh frequency (321 Hz). Besides, the responses corresponding to the direct connection with this component like the test carrier, test sun ($T^D_{2t,ct} \times u_{ct}$ and $T^D_{2t,st} \times u_{st}$) present a higher vibration level.

Another meaningful PPC plot is the phase PPC plot. Figure 7 shows phase PPC plot of the component of the constructed response of the 2nd test planet at (a) 100 Hz and (b) 321 Hz.

At 100 Hz, contributions to the 2nd test planet should only be due to u_{rt}, u_{st}, u_{ct}, u_{1t}, u_{3t}, u_{rr}, those of u_{sr}, u_{cr}, u_{1r}, u_{2r} and u_{3r} being negligible whereas contributions to the 2nd test planet are only due to u_{rt}, u_{st}, u_{ct}, u_{1t}, u_{3t}, u_{3r}, those of u_{sr}, u_{cr}, u_{rr}, u_{1r} and u_{2r} are negligible at 321 Hz.

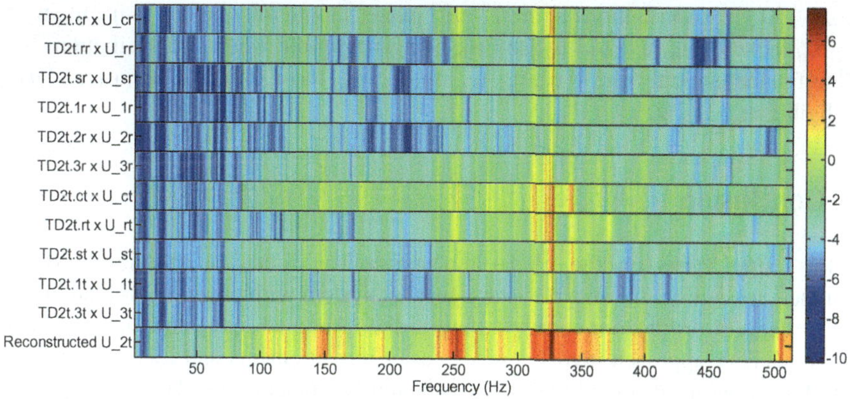

Fig. 6. PPC plot of the decomposition of reconstructed response of the 2nd test planet

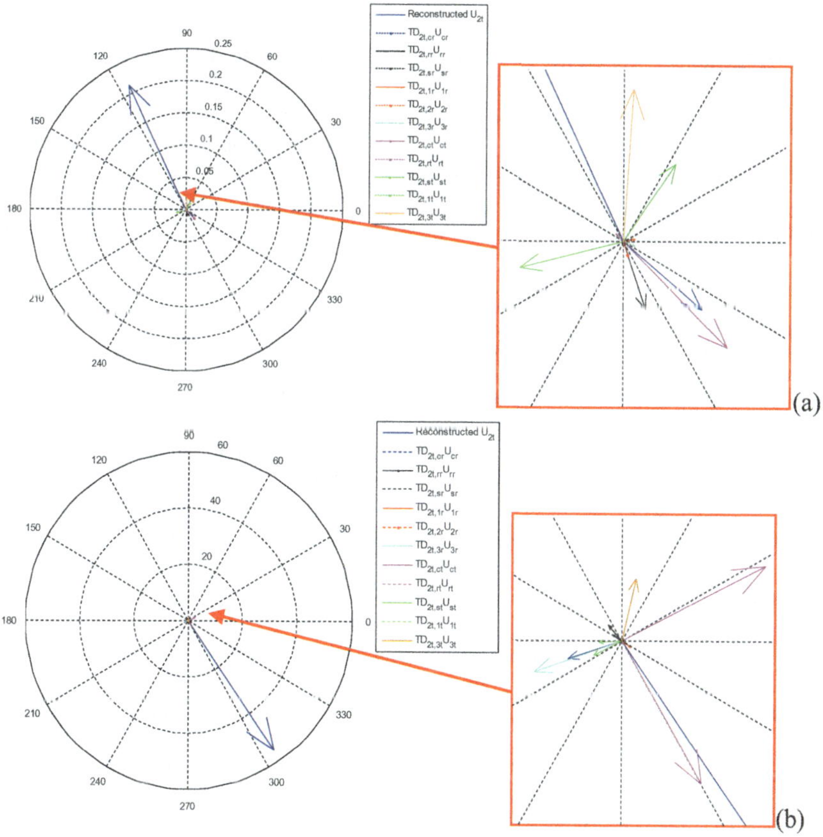

Fig. 7. Phase PPC plot of the constructed response of the 2nd test planet at (a) 100 Hz, (b) 321 Hz

At 100 Hz, contributions to the 2nd test planet should only be due to u_{rt}, u_{st}, u_{ct}, u_{1t}, u_{3t}, u_{rr}, those of u_{sr}, u_{cr}, u_{1r}, u_{2r} and u_{3r} being negligible whereas contributions to the 2nd test planet are only due to u_{rt}, u_{st}, u_{ct}, u_{1t}, u_{3t}, u_{3r}, those of u_{sr}, u_{cr}, u_{rr}, u_{1r} and u_{2r} are negligible at 321 Hz.

4 Conclusion

In this paper, tests on the global transmissibility direct transmissibility (GTDT) approach are carried out on a planetary gear set system with power recirculation where twelve degrees of freedom corresponding to the rotational movement are considered.

The GTDT approach involves much easier measurements and it can be applied to complex cases of practical interest like the studied planetary gearbox.

In this approach, the global transmissibility verified several issues related to the concept of direct transmissibility which allows detailing the response of a system degree-of-freedom in terms of the other system degree-of-freedom and of its response to the external force acting on it.

Acknowledgements. This paper was financially supported by the Tunisian-Spanish Joint Project No. A1/037038/11.

The authors would like also to acknowledge project "Development of methodologies for the simulation and improvement of the dynamic behavior of planetary transmissions DPI2013-44860" funded by the Spanish Ministry of Science and Technology.

Acknowledgment to the University of Cantabria cooperation project for doctoral training of University of Sfax's students.

References

De Sitter, G., Devriendt, C., Guillaume, P., Pruyt, E.: Operational transfer path analysis. Mech. Syst. Signal Process. **24**, 416–431 (2010)

De Klerk, D., Ossipov, A.: Operational transfer path analysis: theory, guidelines and tire noise application. Mech. Syst. Signal Process. **24**, 1950–1962 (2010)

Gajdatsy, P., Janssens, K., Desmet, W., Van der Auweraer, H.: Application of the transmissibility concept in transfer path analysis. Mech. Syst. Signal Process. **24**, 1963–1976 (2010)

Guasch, O., Garcia, C., Jové, J., Artis, P.: Experimental validation of the direct transmissibility approach to classical transfer path analysis on a mechanical setup. Mech. Syst. Signal Process. **37**, 353–369 (2013)

Guasch O., Magrans F.X.: The global transfer direct transfer method applied to a finite simply supported elastic beam. J. Sound Vib. **276**, 335–359 (2004)

Guasch O.: Direct transfer functions and path blocking in a discrete mechanical system. J. Sound Vib. **321**, 854–874 (2009)

Hammami, A., Fernández, A., Viadero, F., Chaari, F., Haddar, M.: Modal analysis of back-to-back planetary gear: experiments and correlation against parameter model. J. Theor. Appl. Mech. **53**(1), 125–138 (2015a) (Warsaw 2015)

Hammami, A., Fernández, A., Viadero, F., Chaari, F., Haddar, M.: Dynamic behaviour of back-to-back planetary gear in run up and run down transient regimes. J. Mech. **31**(4), 48 (2015)

Hammami, A., Fernández, A., Chaari, F., Iglesias, M., Viadero, F., Haddar, M.: Effects of variable loading conditions on the dynamic behaviour of planetary gear with power recirculation. Measurement **94**, 306–315 (2016)

Hammami, A., Iglesias, M., Fernández, A., Chaari, F., Viadero, F., Haddar, M.: Load sharing behavior in planetary gear set. In: Multiphysics Modelling and Simulation for Systems Design and Monitoring Applied Condition Monitoring, vol. 2, pp. 459–468 (2017). https://doi.org/10.1007/978-3-319-14532-7_47

Janssens, K., Gajdatsy, P., Gielen, L., Mas, P., Britte, L., Desmet, W.: OPAX: a new transfer path analysis method based on parametric load models. Mech. Syst. Signal Process. **25**, 1321–1338 (2011)

Maia, N.M.M., Silva, J.M.M., Ribeiro. A.M.R.: The transmissibility concept in multi-degree-of-freedom systems. Mech. Syst. Sig. Proc. **15**, 129–137 (2001)

Margans, F.X.: Method of measuring transmission paths. J. Sound Vib. **74**(3), 321–330 (1981)

Plunt, J.: Finding and fixing vehicle NVH problems with transfer path analysis. Sound Vib., 12–16 (2005)

Ribeiro, A.M.R., Silva, J.M.M., Maia, N.M.M.: On the generalization of the transmissibility concept. Mech. Syst. Signal Process., 29–35 (2000)

Roozen, N.B., Leclère, Q.: On the use of artificial excitation in operational transfer path analysis. Appl. Acoust. **74**, 1167–1174 (2013)

Modeling the Transmission Path Effect in a Planetary Gearbox

Oussama Graja[1(✉)], Bacem Zghal[1], Kajetan Dziedziech[2], Fakher Chaari[1], Adam Jablonski[2], Tomasz Barszcz[2], and Mohamed Haddar[1]

[1] Laboratory of Mechanics, Modeling and Production (LA2MP), National School of Engineers of Sfax (ENIS), Sfax, Tunisia
grajaoussama@gmail.com
[2] Academia Gorniczo Hutnicza (AGH), National Center for Research and Development, Warsaw, Poland

Abstract. In such mechanical systems, as helicopters and self-propelled cranes, designers need to use gearboxes which have an important reduction ratio within compact space. Hence, planetary gearboxes are widely used. Consequently, its monitoring presents an important task for researchers and engineers either in healthy or damaged case. Many researchers are interested on the investigation of the modulation phenomenon in planetary gearbox. It is presented in a frequency representation as side-band activity near to the gear-mesh frequency component and its harmonics. In a healthy case, the origin of this phenomenon in a planetary gearbox (stationary ring) is that the transducer, which is mounted on the external housing of the ring gear, perceived signals from all components including sun-gear, ring-gear, carrier and planet-gears which can occupy different position in one carrier period rotation. Hence, when the planet comes closer to the sensor, the vibration signal increases and vice-versa. In this work, a two dimensional linear lumped parameter model is proposed to model vibration sources. A mathematical formulation of the transmission path is introduced in order to model only the amplitude modulation phenomenon due to the change of the planet-gear position since the speed of the sun is constant. A frequency representation of numerical results is presented and analyzed.

Keywords: Planetary gearbox · Transmission path · Modulation function

1 Introduction

Due to its importance, the amplitude modulation phenomenon occurring in a planetary gearbox is investigated by several researchers to clarify that this phenomenon is a major characteristic in healthy case and differs from a modulation due to a fault. In addition, this phenomenon will influence the overall vibration signal collected by a transducer mounted on the external housing of the gearbox.

Sondkar and Kahraman (2013) proposed a three dimensional lumped parameters model of a double helical planetary gear-set. Firstly, their work aimed at predicting the amplitude of the maximum dynamic mesh force; secondly, it aimed at evaluating the

A. Felkaoui et al. (Eds.): SIGPROMD'2017, ACM 12, pp. 116–122, 2019.
https://doi.org/10.1007/978-3-319-96181-1_11

change of the mesh force under a radially floating sun gear. Inalpolat and Kahraman (2009) developed a mathematical model to report the origin of modulation side-band including in a healthy planetary gear-set. As investigated, the modulation side-band in a frequency representation comes from an epicyclic gearbox having either a stationary sun gear or a stationary ring gear. In a later work (2010), they proposed a nonlinear dynamic model to evaluate the modulation activity in an unhealthy planetary gear-set in the form of run out or eccentricity. Fluctuating mesh force are investigated also in the work of Guo and Parker (2010) and it was related to tooth wedging which causes bearing failures. A model was developed to combine bearing clearance, tooth separation and wedging and back-side contact. The modulation phenomenon either amplitude or frequency modulation in time domain which is called side-band activity in a frequency representation were investigated in the work of Feng and Zuo (2012), Liang et al. (2015) and Liu et al. (2016). In Feng and Zuo (2012), the authors simulate faulty gear damages for instance faulty planet gear and faulty sun gear after defining characteristic frequency of faulty gear in a planetary gearbox. In Liang et al. (2015), a lumped parameter model was developed to build vibration sources then all vibration was concluded in the sensor location by taking into account the transmission path effect due to the rotational motion of the carrier which holds planet gears. Two vibration properties were investigated: healthy case and cracked tooth case. In a later work Liu et al. (2016), they focused only on transmission path which is modeled as two parts: a first part inside the gearbox to the housing and a second part along the housing to the sensor location.

This paper is organized as follows: In Sect. 2, the origin of the modulation phenomenon is investigated and the transmission path is defined. In Sect. 3, the transmission path is formulated as function of geometric and physical parameters of the planetary gearbox. Finally, some numerical results are presented in Sect. 4 where the impact of planets on the resultant vibration is investigated and the vibration characteristics are revealed.

2 Origin of the Modulation Phenomenon

As mentioned in Sect. 1, one planet can occupy different positions in one carrier rotation period. Figure 1 presents three locations of one planet. The sensor is mounted on the external housing of the planetary gearbox. It can acquire signals due to the vibration coming from all components including sun-gear, ring-gear, carrier and planet-gear. All components have only a rotational motion with respect to its center except the planet-gear which has an additional motion with respect to the center of the gearbox. Due to this additional motion, the vibration signals are under modulation phenomenon either amplitude modulation in case of stationary speed or amplitude and frequency modulation in case of fluctuating speed.

Since the study is focused on the stationary conditions, we will investigate only the amplitude modulation phenomenon. As shown in Fig. 1, the transmission path can be divided into two parts: a first part inside the gearbox (blue one) and a second part along the casing. As the planet moves, the dimension of the blue path is still constant. On the

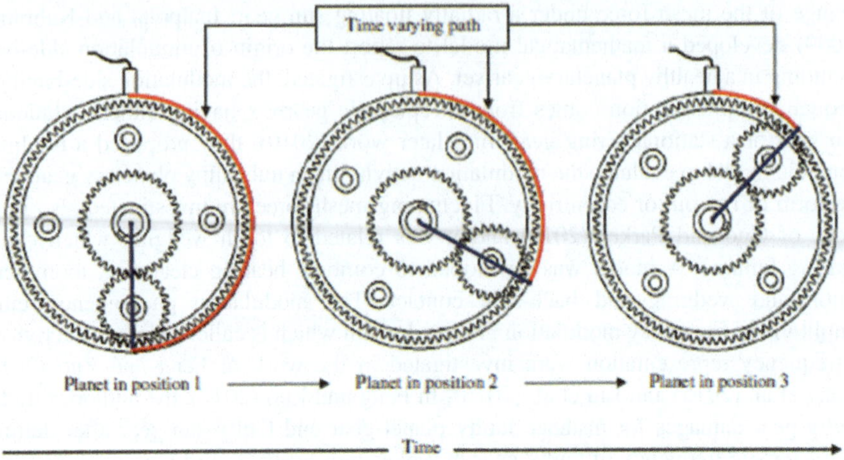

Fig. 1. Different position of one planet

other hand, the dimension of the red path decreases which creates the amplitude modulation function (AMF). Hence, two points have to be mentioned:

- the main cause of the amplitude modulation is the time varying transmission path which is represented by the arc of the circle.
- when the dimension of the path decreases, the vibration signal increases since the planet becomes closer to the transducer. Therefore, the AMF and the time varying path are inversely proportional.

3 Mathematical Formulation of the Transmission Path

In Sect. 2, it was mentioned that the time varying path (origin of modulation phenomenon) is an arc of the circle. Hence, a geometric construction presented in Fig. 2(a) was made in order to link the time varying arc with geometric parameters of the planetary gear-set.

Derived from the geometrical construction given in Fig. 2(a), the AMF can be expressed as:

$$AMF = \frac{\frac{1}{2 \times R_r \times \sin(Arc/R_r) + 1}}{\max\left(\frac{1}{2 \times R_r \times \sin(Arc/R_r) + 1}\right)} \tag{1}$$

An offset equal to one is taken into account to avoid the division by zero since the AMF and the time varying transmission path are inversely proportional. In addition, the AMF is divided by its max to consider only the percentage of the function.

Figure 2(b) turns out the shape of the AMF mentioned in Eq. 1.

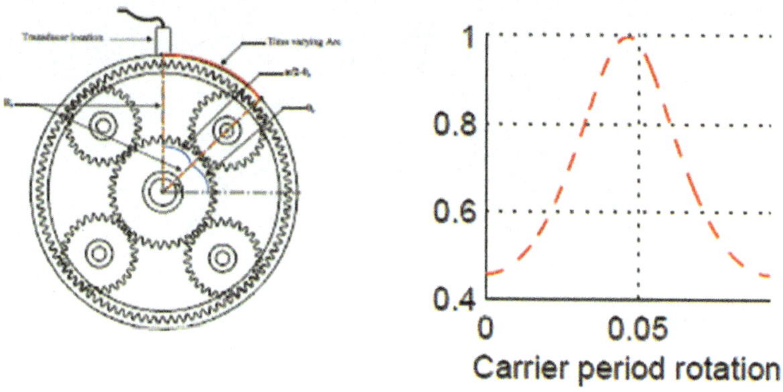

(a) (b)

Fig. 2. (a) Geometric construction (b) Modulation function

4 Numerical Simulation

Table 1 resumes physical parameters of the planetary gear-set to simulate its dynamic behavior. The dynamic model used is shown in Fig. 3. Acceleration is measured with respect to the carrier.

Table 1. Physical parameters of a planetary gear set

Parameters	Sun gear	Planet gear	Ring gear	Carrier
Number	1	4	1	1
Number of teeth	39	27	93	–
Modulus	2	2	2	–
Pressure angle	20	20	20	–
Mass (Kg)	2.3	0.885	2.94	15
Base circle radius (m)	0.078	0.054	0.186	0.132
Bearing stiffness	$K_{sx} = K_{sy} = K_{px} = K_{py} = K_{rx} = K_{ry} = K_{cx} = K_{cy} = 10^8$			
	$K_{sw} = 0$	$K_{pw} = 0$	$K_{rw} = 10^{15}$	$K_{cw} = 0$
Input torque (Nm)	150	–	–	–
Input-speed (tr/mn)	2183.6	–	–	–

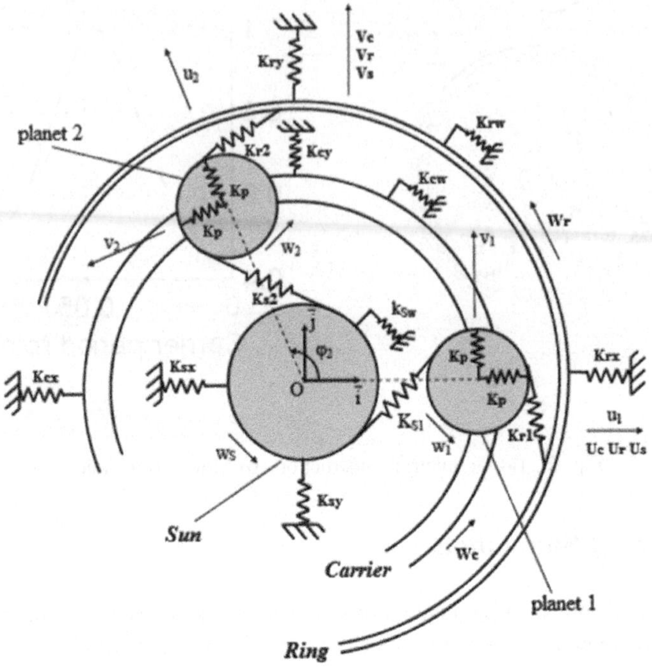

Fig. 3. Lumped parameters model

4.1 Impact of Each Planet on the Resultant Vibration

Figure 4 highlights the contribution of each planet alone on the overall resultant vibration. As shown, the consequence of the passage of each planet is presented by the modulation of the vibration collected by the transducer.

4.2 Analysis of Numerical Results

Figure 5 displays a zoom section between 2500 Hz and 4500 Hz of the frequency representation of the resultant vibration. As seen, there is a side-band activity near to two gear-mesh frequency harmonics (H3GMF = 3000 Hz and H4GMF = 4000 Hz). To identify the origin of side-band components, it is necessary to calculate the carrier frequency.

Based on Table 1 and on formulas presented below, the frequency of the carrier is obtained.

$$f_s = \frac{N_s}{60} = 36.39\,\text{Hz} : \textit{sun frequency};\ r = \frac{Z_s}{Z_s + Z_r} = 0.2955 : \textit{ratio (planetary)}$$

$$f_c = r \times f_s = 10.75\,\text{Hz} : \textit{carrier frequency}$$

Table 2 resumes the identified frequencies in Fig. 5 and its correspondences.

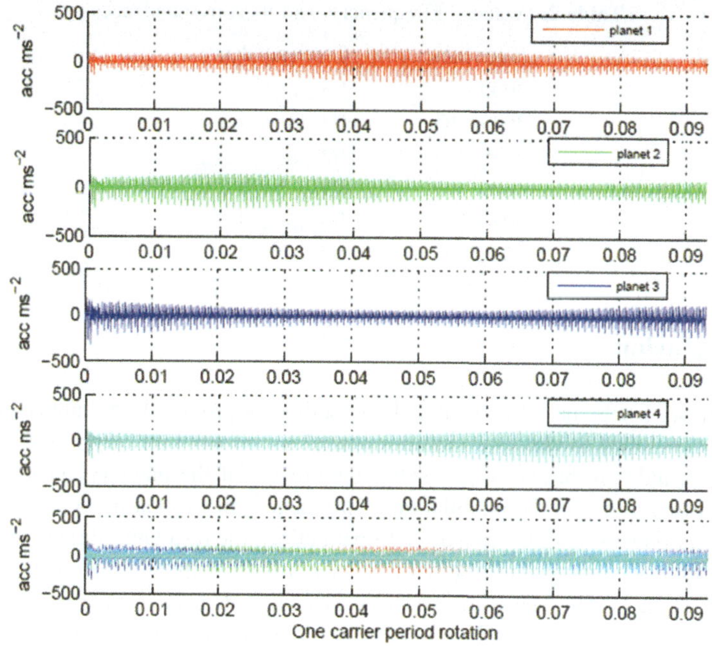

Fig. 4. Contribution of each planet on the resultant vibration

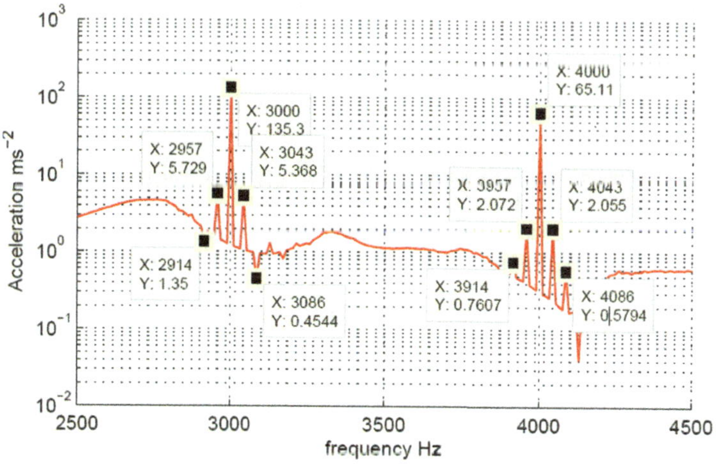

Fig. 5. Frequency representation of the resultant vibration

Table 2. Values of frequencies and its correspondences

Frequencies (Hz)	Correspondence
3000	$H3_{GMF}$
3043–2957	$H3_{GMF} \pm 4 \times f_c$
3086–2914	$H3_{GMF} \pm 8 \times f_c$
4000	$H4_{GMF}$
4043–3957	$H4_{GMF} \pm 4 \times f_c$
4086–3914	$H4_{GMF} \pm 8 \times f_c$

5 Conclusion

In this study, a vibration signal modeling method is proposed. A lumped parameter model is developed to simulate vibration signal issued from all components. A mathematical formulation based on geometric parameters of the planetary gearbox is presented in order to model the modulation phenomenon. Incorporating vibrations coming from all components and the transmission path effect, the resultant vibration is obtained at the sensor location as the sum of all vibration components influenced by the transmission path. The spectrum structure is analyzed and the side-band activity is predicted near to the gear-mesh frequency components and its harmonics.

Acknowledgements. This work is partially supported by NATIONAL SCHOOL OF ENGINEERS OF SFAX (ENIS)/Laboratory of Mechanics, Modeling and Production (LA2MP) and the National Centre of Research and Development (NCRD) in Poland under the research project no. PBS3/B6/21/2015.

References

Sondkar, P., Kahraman, A.: A dynamic model of a double-helical planetary gear set. Mech. Mach. Theory **70**, 157–174 (2013)

Inalpolat, M., Kahraman, A.: A theoretical and experimental investigation of modulation sidebands of planetary gear sets. J. Sound Vib. **323**, 677–696 (2009)

Inalpolat, M., Kahraman, A.: A dynamic model to predict modulation sidebands of a planetary gear set having manufacturing errors. J. Sound Vib. **329**, 371–393 (2010)

Guo, Y., Parker, R.G.: Dynamic modeling and analysis of a spur planetary gear involving tooth wedging and bearing clearance nonlinearity. Eur. J. Mech. A/Solids **29**, 1022–1033 (2010)

Feng, Z., Zuo, M.J.: Vibration signal models for fault diagnosis of planetary gearboxes. J. Sound Vib. **331**, 4919–4939 (2012)

Liang, X., Zuo, M.J., Hoseini, M.R.: Vibration signal modeling of a planetary gear set for tooth crack detection. Eng. Fail. Anal. **48**, 185–200 (2015)

Liu, L., Liang, X., Zuo, M.J.: Vibration signal modeling of a planetary gear set with transmission path effect analysis. Measurement **85**, 20–31 (2016)

Dynamic Behavior of Spur Gearbox with Elastic Coupling in the Presence of Eccentricity Defect Under Acyclism Regime

Atef Hmida(✉), Ahmed Hammami, Fakher Chaari, Mohamed Taoufik Khabou, and Mohamed Haddar

Laboratory of Mechanics, Modeling and Production (LA2MP), National School of Engineers of Sfax, BP 1173, 3038 Sfax, Tunisia
abdallahatef@yahoo.fr, ahmed.hammami2109@gmail.com, fakher.chaari@gmail.com, mtkhabou@hotmail.com, mohamed.haddar@enis.rnu.tn

Abstract. In this paper, the effect of eccentricity defect on the dynamic behaviour of one stage spur gearbox running under acyclism regime is studied. In fact, acyclism regime is generated by a combustion engine motor which produced fluctuations of load and speed. The motor torque is periodic and it modeled in the force's vector. The rotational speed of the Diesel engine is a harmonic function and it generates a periodic fluctuation of the gear meshing stiffness function. This driven motor is joined to the gearbox through an elastic coupling in which the model of Nelson and Crandall is adopted. The eccentricity defect is introduced in the pinion. This defect produces an additional potential energies and kinetic energy and it is modelled through additional forces. The equation of motion is obtained using Lagrange formalism and the algorithm of Newmark is used to compute the dynamic response of the studied system and the Wigner–Ville distribution shows the dynamic behaviour of the gearbox under this cyclo-stationary regime. Results show the variability of the meshing frequency and its harmonics which excites the system. Also, natural frequencies are observed in the spectrum and Wigner–Ville distribution of the dynamic signal. Nevertheless, these methods fail to detect the frequencies of eccentricity and acyclism.

Keywords: Eccentricity · Acyclism · Elastic coupling · Spur gearbox

1 Introduction

Acyclism is a transient regime. It is generated by a combustion engine and it is characterized by fluctuations of speed and torque.

Many researchers focused on this regime: Barthod et al. (2007a) studied the effect of acyclism on the rattle threshold inside different gearbox configurations. Sika and Velex (2008) used a torsional gear model to study the effect of engine speed fluctuations which is considered as a sinusoidal and multi-harmonic function. Khabou et al. (2011) investigated a spur motored by a diesel engine where its applied torque is considered as a multi-harmonic function.

A. Felkaoui et al. (Eds.): SIGPROMD'2017, ACM 12, pp. 123–132, 2019.
https://doi.org/10.1007/978-3-319-96181-1_12

In addition to the motor regime, defects are source of excitation. For example, the eccentricity which is due to the non-concentricity between the axis of the pitch cylinder of the gear and the axis of rotation of the shaft is investigated by many researchers:

Driss et al. (2014) studied the dynamic behavior of two-stage straight bevel gear with some defects which are the eccentricity defects, profile error and cracked tooth. They proposed a new method for modeling gear mesh stiffness of straight bevel gear and they introduced these defects in the three dimensional model of the studied system. As results, the spectrum dynamic response shows appearance of sidebands around the meshing frequency excited by the fault and its harmonics. Walha et al. (2011) studied the effect of eccentricity defect on the dynamic behavior of an automotive clutch coupled with a two stage helical gear. They included also three types of nonlinearity which are dry friction path, double stage stiffness and spline clearance. Chaari et al. (2006) studied the influence of eccentricity on the sun gear on the dynamic behavior of planetary gear. They introduced this defect by adding a transmission error modeled as displacement on the line of action of the sun-planet gearmesh.

In this work, effects of acyclism and eccentricity defect on the dynamic behavior of a spur gearbox with elastic coupling are investigated. The system is powered by diesel engine and the second model of Nelson and Crandall (1992) is adopted for the elastic coupling. Excitations due to the fluctuation of load and speed of acyclism regime and eccentricity defect of gearbox are introduced to the dynamic model. The dynamic response is computed through Newmark algorithm and results are shown using Wigner–Ville distributions.

2 Dynamic Model

The studied system is composed by a diesel engine motor and a receiver which are connected through one stage of spur gearbox and an elastic coupling located between the motor and the pinion.

Figure 1 show the corresponding dynamic model which is divided into three blocks. This model was proposed by Hmida et al. (2016, 2017). The pinion and the wheel of gearbox and the diesel engine motor are assumed as rigid bodies. Transmission shafts are assumed massless and have torsional stiffness $K\theta i$ and torsional damping $C\theta i$ (i = 1, 2, 3). They are supported by bearings which are modeled with parallel springs (Kxi, Kyi) and damping (Cxi, Cyi).

The model of Nelson and Crandall (1992) is adopted for the elastic coupling because this model is best approach to describe the dynamics of elastic couplings (Tadeo and Cavalca 2003; Tadeo et al. 2011). This coupling is modeled with two translation stiffness (K_{xc}, K_{yc}), a torsional stiffness ($K_{\theta c}$), two translation damping (C_{xc}, C_{yc}) and a torsional damping ($C_{\theta c}$). Its inertial effects are included in the first block (I_{12}) and the second blocks (I_{21}).

The degree of freedom vector "q" is defined as following:

$$q = (\theta_{11}, \theta_{12}, \theta_{21}, \theta_{22}, \theta_{31}, \theta_{32}, x_1, y_1, x_2, y_2, x_3, y_3) \tag{1}$$

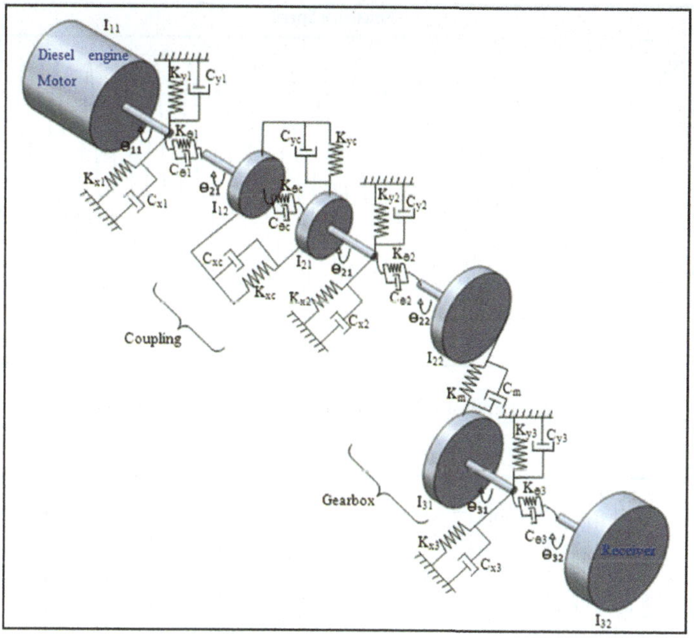

Fig. 1. Dynamic model of spur gearbox with an elastic coupling

2.1 Acyclism Modeling

During the power stroke, the diesel engine generates a variable speed and torque.

The rotational speed of the engine $\Omega(t)$ written by Sika and Velex (2008) as following:

$$\Omega(t) = \Omega_{10}\left(1 + \sum\nolimits_n \rho_n(\Omega_{10}) \sin(n\Omega_{10}t + \varphi_n)\right) \tag{2}$$

where Ω_{10} is the average velocity.

n is the harmonic of the generated speed function, ρ_n and φ_n are respectively the corresponding amplitude and phase. According to the Eq. (2) limited on the 1st harmonic, the evolution of the rotational speed generated by the diesel engine motor is shown in Fig. 2.

The shape of the rotational speed of the Diesel engine generates a periodic fluctuation of the gear meshing stiffness function (K_m) as shown in Fig. 3.

According to Ligier and Baron (2002), the torque C_m developed by the combustion engine can be written as:

$$C_m \approx \overline{C_m} + \frac{P_{max}}{192} V_{cyl}(0.46 \sin 2\alpha_c + 0.24 \sin 4\alpha_c + 0.03 \sin 6\alpha_c) \tag{3}$$

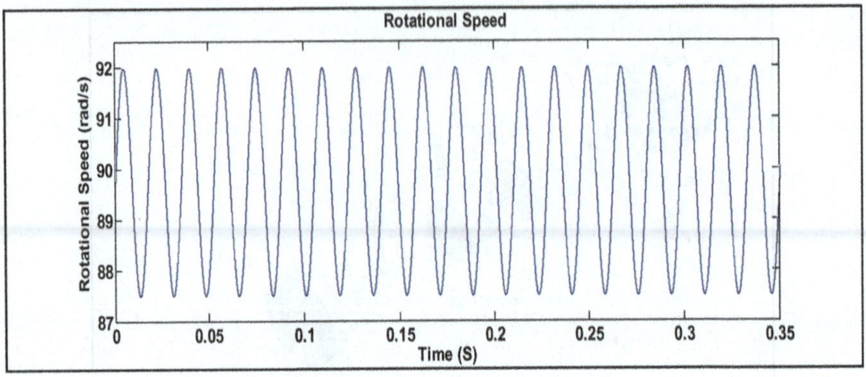

Fig. 2. Time evolution of the engine rotational speed

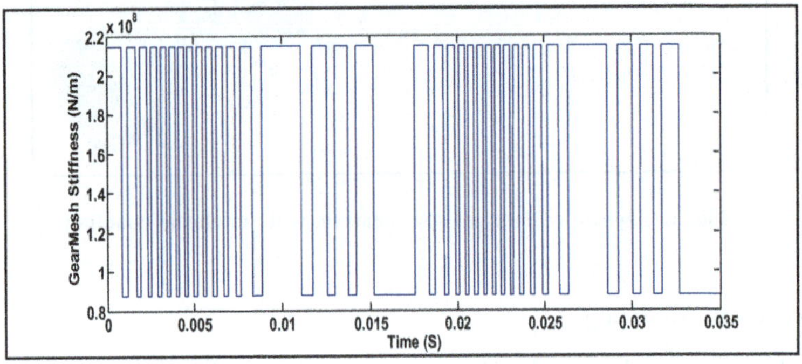

Fig. 3. Time evolution of the gear mesh stiffness

Where $\overline{C_m}$ and α_c are respectively the average of engine torque and the angular position of the crankshaft. V_{cyl} and P_{max} are respectively the cylinders capacity and the maximum pressure inside cylinders.

The applied torque is periodic and it is shown in Fig. 4.

2.2 Eccentricity Modeling

The approach of modelling of the eccentricity is based on an eccentricity error due to a deviation between the center of rotation of the gear and its geometric center. Michalec (1966) considered the case of a single eccentric gear and showed that the transmission kinematic error was a deterministic perturbation of frequency f_d the frequency of defect which is equal to the frequency of rotation f_r of the pinion. The amplitude is proportional to its eccentricity.

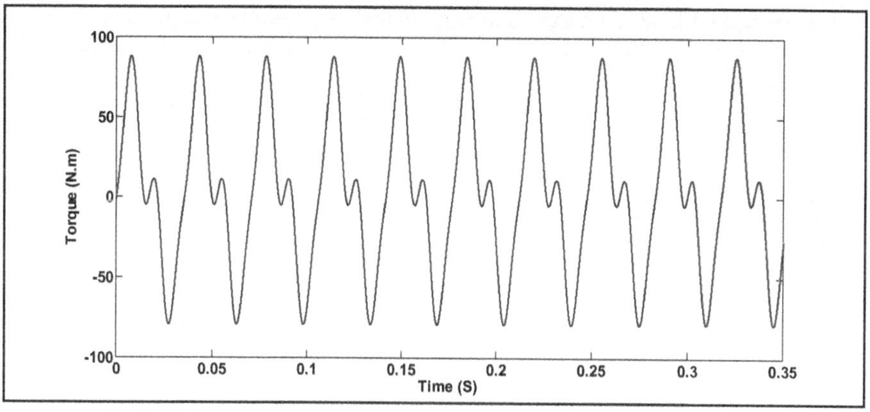

Fig. 4. Time evolution of the engine torque

E_{22} is the distance between the axis of rotation and the axis of inertia of the wheel and expressed by:

$$e_{22}(t) = e_{22}\sin(\Omega_{22}t - \lambda_{22}) \tag{4}$$

Where $\Omega_{22} = 2\pi f_d$ (f_d: frequency of defect)

e_{22} and λ_{22} are respectively the amplitude of eccentricity and the phase of eccentricity. They are shown in Fig. 5.

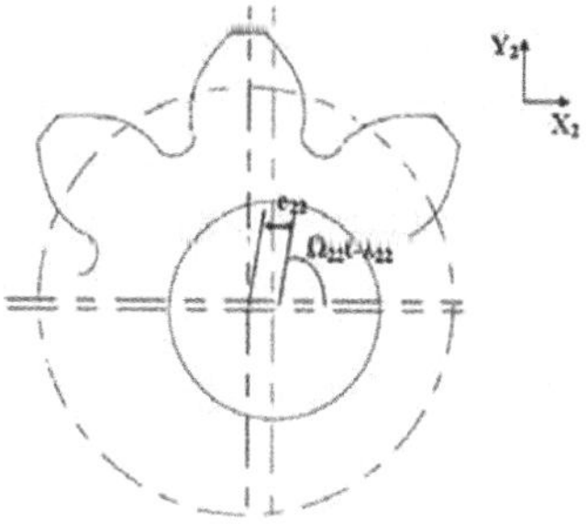

Fig. 5. Eccentricity defect

The eccentricity defect affects the potential energies and kinetic energy. In fact, this defect affects the tooth deflections. So, there is an additional potential energy which is modelled by an additional force:

$$F^{p}_{ecc} = K_m(t)e_{12}(t)\{0 \quad 0 \quad 0 \quad r_{b21} \quad r_{b22} \quad 0 \quad 0 \quad 0 \quad \sin(\alpha) \quad \cos(\alpha) \quad -\sin(\alpha) \quad -\cos(\alpha)\} \tag{5}$$

The additional kinetic energy is also modelled as an additional force:

$$F^K_{ecc} = -m_{22}e_{22}\Omega^2_{22}\{0 \quad 0 \quad 0 \quad 0 \quad 0 \quad 0 \quad 0 \quad 0 \quad \cos(\Omega_{22}t - \lambda_{22}) \quad \sin(\Omega_{22}t - \lambda_{22}) \quad 0 \quad 0\} \quad (6)$$

3 Equation of Motion

The equation of motion is obtained using Lagrange formalism:

$$[M]\ddot{q} + ([C_m] + [C_s])\dot{q} + ([K(t)] + [K_s])q = F(t) \quad (7)$$

[M] is the global mass matrix. $[K_s]$ and [K(t)] are respectively the structural stiffness matrix of the system and the time varying mesh stiffness matrix. $[C_s]$ and $[C_m]$ are respectively the structural damping matrix and the mesh damping matrix.

The external force vector [F(t)] is defined as:.

$$F(t) = F^p_{ecc}(t) + F^K_{ecc}(t) + F_{ext}(t) \quad (8)$$

$$F_{ext} = \{C_m \quad 0 \quad 0 \quad 0 \quad 0 \quad -C_r \quad 0 \quad 0 \quad 0 \quad 0 \quad 0 \quad 0\} \quad (9)$$

C_m and C_r are respectively the motor torque and the receiver torque.

These entire matrixes are defined in Hmida et al. (2016).

4 Numerical Results

In this part, effect of acyclism and eccentricity are induced in the model and numerical results are carried out using the parameters values of the dynamic model presented in Table 1.

Newmark method is used to compute the numerical results.

Figure 6 shows the time displacement signal of the second bloc in the Y direction. The observed fluctuations on this figure correspond to the influence of the meshing and eccentricity phenomena on the dynamic response.

Time acceleration signal of the second bloc in the X direction is shown in Fig. 7. This signal is modulated by the acyclism and the eccentricity defect.

Spectral analysis is the most widely used techniques. Indeed, analysis spectrum of acceleration of the second bloc in the X direction (Fig. 8) shows several peaks in the neighborhoods of the natural frequencies of the system f_i which are resumed in Table 2.

Table 1. Values of the model parameters

Gear box parameters	
Teeth number	$Z_{12} = 20$; $Z_{21} = 50$
Mass (Kg)	$m_{12} = 1.77$; $m_{21} = 2.5$
Pressure angle	$\alpha = 20°$
Teeth module (m)	$m_n = 2 \times 10^{-3}$
Contact ratio	$\varepsilon_\alpha = 1.6$
Average mesh stiffness (N/m)	$K_{moy} = 2.11 \times 10^8$
Coupling's characteristics	
Inertia (Kg m^2)	4×10^{-3}
Mass (Kg)	4.5
Torsional stiffness (Nm/rad)	352
Translation stiffness (N/m)	462×10^2
Engine motor's characteristics	
Inertia (Kg m^2)	4×10^{-3}
Maximum pressure inside cylinders P_{max} (Bar)	49
Average of engine torque $\overline{C_m}$ (N m)	17.5
Cylinders capacity Vcyl (cm^3)	2000
Receiver's characteristics	
Inertia (Kg m^2)	6×10^{-3}
Characteristics of shafts and bearings	
Torsional Shaft stiffness (Nm/rad)	5×10^5
Bearing stiffness (N/m)	5×10^8
Characteristics of eccentricity	
Amplitude of eccentricity (μm)	50
Phase of eccentricity (rad)	$\pi/6$

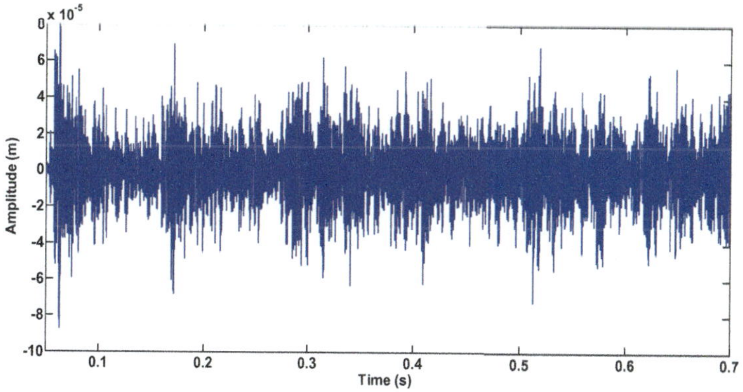

Fig. 6. Time displacement signal of the second bloc in the X direction

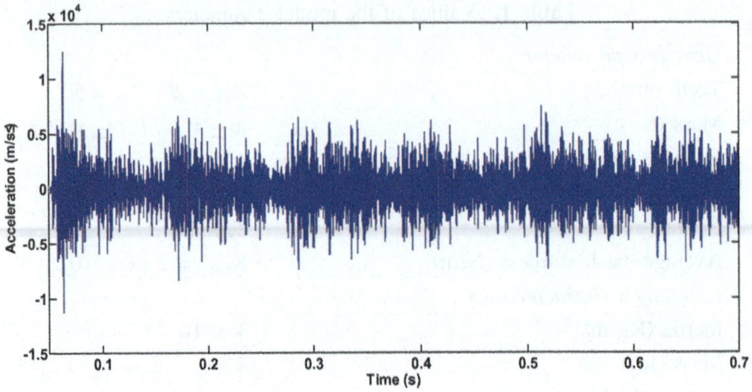

Fig. 7. Time acceleration signal of the second bloc in the X direction

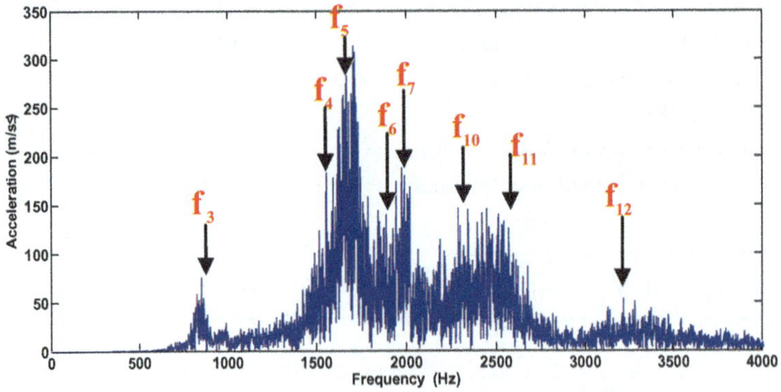

Fig. 8. The spectrum of acceleration of the second bloc in the X direction

Table 2. The natural frequencies

Natural freq	f_1	f_2	f_3	f_4	f_5	f_6	f_7	f_8	f_9	f_{10}	f_{11}	f_{12}
Hz	0	26	891	1583	1664	1867	1902	1974	1974	2344	2459	3292

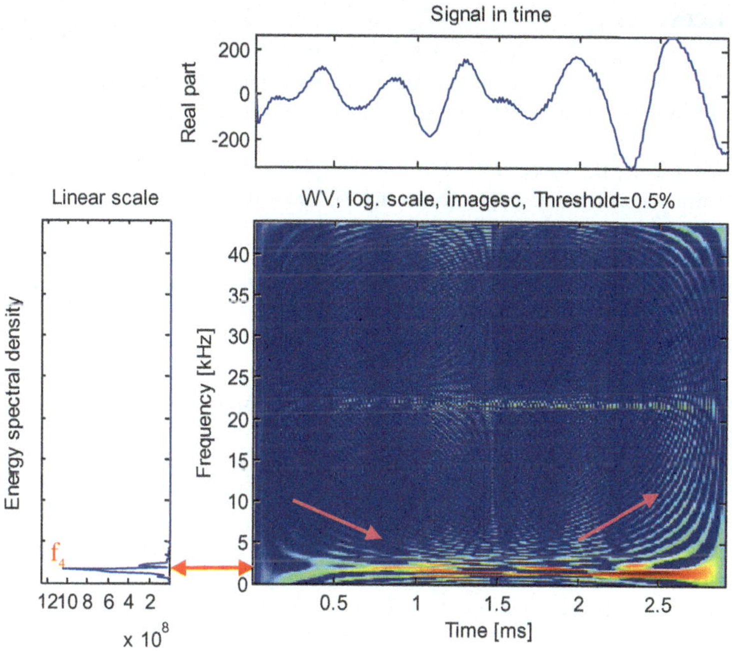

Fig. 9. Wigner–Ville distribution of the acceleration in direction X2

Figure 9 represents the Wigner–Ville distribution of the acceleration of the second bloc in the X direction in order to analyze the non-stationary behavior of the signal.

Due to the acyclism regime, it can be seen the variability of the meshing frequency and its harmonics which excites the system, and horizontal lines represented peaks in the natural frequencies of the system. Nevertheless, the frequency of acyclism and the frequency of eccentricity defect are not observed.

5 Conclusion

In this paper, the dynamic behavior of spur gearbox with an elastic coupling is studied under acyclism regime generated by combustion engine. The eccentricity defect is introduced in the model. Spectral analysis and the Wigner–Ville distribution of the dynamic response are used to provide information about their state. In fact, only peaks in the natural frequencies of the system and the meshing frequencies appear. These methods fail to detect the frequencies of eccentricity and acyclism.

The future study will mainly focus on the decomposition of the dynamic response of the system with eccentricity under acyclism regime in order to extract the frequencies of excitations from the non-stationary signals.

References

Barthod, M., Hayne, B., Tebec, J.-L., Pin, J.-C.: Experimental study of dynamic and noise produced by a gearing excited by a multi-harmonic excitation. Appl. Acoust. **68**(9), 982–1002 (2007a)

Barthod, M., Hayne, B., Tebec, J.-L., Pin, J.-C.: Experimental study of gear rattle excited by a multi-harmonic excitation. Appl. Acoust. **68**(9), 1003–1025 (2007b)

Chaari, F., Fakhfakh, T., Hbaieb, R., Louati, J., Haddar, M.: Influence of manufacturing errors on the dynamic behavior of planetary gears. Int. J. Adv. Manuf. Technol. **27**, 738–746 (2006)

Driss, Y., Hammami, A., Walha, L., Haddar, M.: Effects of gear mesh fluctuation and defaults on the dynamic behavior of two-stage straight bevel system. Mech. Mach. Theory **82**, 71–86 (2014)

Hmida, A., Hammami, A., Khabou, M.-T., Chaari, F., Haddar, M.: Modal analysis of spur gearbox with an elastic coupling. Adv. Acoust. Vib. Appl. Condition Monit. **5**, 153–163 (2016). https://doi.org/10.1007/978-3-319-41459-1_15

Hmida, A., Hammami, A., Khabou, M.T., Chaari, F., Haddar, M.: Effect of elastic coupling on the modal characteristics of spur gearbox system. Appl. Acoust. (2017). https://doi.org/10.1016/j.apacoust.2017.06.013

Khabou, M.-T., Bouchaala, N., Chaari, F., Fakhfakh, T., Haddar, M.: Study of a spur gear dynamic behavior in transient regime. Mech. Syst. Signal Process. **25**(8), 3089–3101 (2011)

Ligier, J.-L., Baron, E.: Acyclisme et vibrations. In: Applications Aux Moteurs Thermiques et Aux Transmissions, vol. 1, Editions TECHNIP, Paris (2002)

Michalec, G.W.: Precision Gearing: Theory and Practice. Wiley and Sons, New York (1966)

Nelson, H.D., Crandall, S.H.: Analytic prediction of rotordynamic response. In: Ehrich, F.E. (ed.) Handbook of Rotordynamics. McGraw-Hill Inc, New York (1992)

Sika, G., Velex, P.: Instability analysis in oscillators with velocity-modulated time-varying stiffness—applications to gears submitted to engine speed fluctuations. J. Sound Vib. **318**:166–175 (2008)

Tadeo, A.T., Cavalca, K.L.: A comparison of flexible coupling models for updating in rotating machinery response. J. Braz. Soc. Mech. Sci. Eng. **25**(3), 235–246 (2003)

Tadeo, A.-T., Cavalca, K.-L., Brennan, M.-J.: Dynamic characterization of a mechanical coupling for a rotating shaft. Proc. Inst. Mech. Eng. Part C J. Mech. Eng. Sci. **225**(3), 604–616 (2011)

Walha, L., Driss, Y., Khabou, M.-T., Fakhfakh, T., Haddar, M.: Effects of eccentricity defect on the nonlinear dynamic behavior of the mechanism clutch-helical two stage gear. Mech. Mach. Theory **46**(7), 986–997 (2011)

Author Index

A
Abbes, Mohamed Slim, 96
Abdennadher, Moez, 87

B
Barszcz, Tomasz, 116
Baslamisli, S. Caglar, 79, 96
Ben Hassen, Dorra, 96

C
Chaari, Fakher, 56, 79, 96, 104, 116, 123

D
Dziedziech, Kajetan, 116

F
Fedala, Semcheddine, 1
Fedala, Semchedine, 16
Felkaoui, Ahmed, 1, 16, 34, 56, 68
Fenineche, Hocine, 34
Fernandez del Rincon, Alfonso, 104

G
Ghorbel, Ahmed, 87
Graja, Oussama, 116

H
Haddar, Maroua, 79
Haddar, Mohamed, 56, 79, 87, 96, 104, 116, 123
Hammami, Ahmed, 104, 123
Hmida, Atef, 123

J
Jablonski, Adam, 116

K
Khabou, Mohamed Taoufik, 123

M
Mahgoun, Hafida, 1, 44, 56
Miladi, Mariem, 96

R
Rémond, Didier, 16
Rezig, Ali, 34

S
Sedira, Miloud, 68
Selmani, Houssem, 16

V
Viadero Rueda, Fernando, 104

W
Walha, Lassâad, 87

Z
Zghal, Becem, 87, 116
Ziani, Ridha, 1, 44, 68

A. Felkaoui et al. (Eds.): SIGPROMD'2017, ACM 12, pp. 133–133, 2019.
https://doi.org/10.1007/978-3-319-96181-1

The manufacturer's authorised representative in the EU is Springer Nature Customer Service Centre GmbH, Europaplatz 3, 69115 Heidelberg, Germany. If you have any concerns regarding our products, please contact ProductSafety@springernature.com

Printed and bound by CPI Group (UK) Ltd, Croydon, CR0 4YY
05/10/2026
02228152-0008